Faith and Physics

Faith and Physics

The Scientific Basis For Spiritual Belief

J. P. Befumo

ISBN #978-0-9767489-2-2 Softcover

This book was printed in the United States of America.

To order additional copies of this book, contact:
Pleasant Mount Press, Inc.
www.PleasantMountPress.com

This book I dedicate to all those who have seen through the veil of illusion and experienced the texture of reality.

Contents

Contents (continued)

Preface

The first issue of *Omni Magazine*[1] was published in October 1978, and I picked it up more or less at random from a New York Subway news stand. Little did I know how it would change my life. At the time I was a high-school dropout, working as longshoreman at Pier 11, in the Red Hook section of Brooklyn, New York. I read it, cover to cover, which I rarely did at that time, and did so several times over. It all sounded so . . . convincing, but could it possibly be *real*? Next payday I stopped at a local university bookstore and picked up a copy of *Quantum Physics of Atoms, Molecules, Solids, Nuclei, and Particles* by Robert Eisberg and Robert Resnick, determined to teach myself enough quantum theory to make an informed judgment. Of course, I had no concept of just how ludicrous this ambition was, in view of my failure to have completed any high school math past sophomore geometry (and that in summer school). I got as far into the preface as where it said: "No more than a knowledge of college calculus is assumed. . . ." Another stop at the book store scored me a copy of *Analytical Geometry and the Calculus*, by A.W. Goodman. It didn't take me long to realize that

[1]Omni Magazine was the *Wired Magazine* of the 70s & 80s. Everyone from Nobel Prize winners to UFO crackpots were interviewed. News of new and fascinating inventions and ideas were a regular part of the magazine. The first issue was published in October 1978, the last in Winter 1995, with an internet version lasting until 1998. Bob Guccione described the magazine in its first issue as "an original if not controversial mixture of science fact, fiction, fantasy and the paranormal" [Guccione, B. First word., *Omni* 1(1):6, October 1978].

teaching myself calculus, so I could teach myself quantum mechanics, was probably not going to happen. So, I took my High School Equivalency exam, and enrolled full time at Brooklyn College in January of 1979, declaring as majors, physics and mathematics.

The rigors of the undergraduate physics curriculum were such that any ideas of searching for the spiritual nature of life in the equations of science were soon relegated to a distant back burner, until I began working as an assistant at the University's 7.5 MeV proton accelerator lab. Part of my services involved checking over the data and calculations for some of the professors, and in this capacity I came face to face with the indisputable evidence of relativistic time dilation. Still, with the effort of my classes, the more metaphysical implications of these observations were filed away for later analysis.

After college, I went to work at Digital Equipment Corporation's Advanced Semiconductor Development Lab in Hudson, Massachusetts, where I was involved in research dealing with quantum anomalies that had begun to reveal themselves in the ultra thin-film circuits being developed at that site. This is where the memory of my original purpose began to seep back into my subconscious.

I left DEC in 1992, so I guess I've been pondering this issue for some fifteen years now — almost double that if we go back to that seminal issue of Omni Magazine. I guess that's not so long for connecting with the nature of spirit and the meaning of life, though it has seemed a rather lengthy path at times.

The substance of reality is not something that can be conveyed with any language; it must be experienced firsthand, and I am convinced that absolute, unshakable belief is a prerequisite for attaining that experience, or gnosis[2]. Sadly, society would often have us feel silly for believing in anything beyond the mundane world reflected (inaccurately) by our physical senses. Hence, my approach will be to attempt to facilitate this unassailable level of belief by showing how the notion of a spiritual world is fully consistent, indeed, is demanded by, the implications of all of our major theories of modern physics.

[2]The word gnosis (from the Greek word for knowledge, γνώσις) refers to a form of spiritual knowledge that is more commonly familiar to people as enlightenment. It is the direct apprehension of the face of reality, and can never be conveyed in words, except in the most metaphorical fashion.

To this end, the first section of this book will concentrate on providing the non-scientist with an intuitive understanding of the relevant scientific theories, concentrating mainly on the two most important theories of modern physics: Einstein's theories of Relativity, and Quantum Theory; and more significantly, on the esoteric implications of these mind-bending concepts. In the second section we'll illustrate some of the underlying conclusions of section I in a way that will appeal to the intuition. In the third section we will develop these concepts to prove, unequivocally, that the spirit exists, that it is eternal, and to describe fully the nature and meaning of life. In part IV, we'll take a leap of faith to see how science provides plenty of room for concepts like Deity and soul. Finally, we'll look at a smattering of spiritual, paranormal, and mystical concepts, practices, and theories, examining them in light of the world models hitherto developed.

The result will be an ability to look at the world with an open mind, and an entirely unprecedented degree of conviction.

The History of this Book

Some time in the late 1990s, these years of dormant cogitations coalesced into what can best be termed visions of the nature of spirit, which I subsequently came to understand as a kind of gnosis, I felt compelled to record, sort, and interrelate these images, mainly to help me make some sense of them for myself, and see if the resulting logic represented a truly apodictic[3] chain of reasoning.

I worked on these more or less full time, through numerous rewrites, until 2005, when the result coalesced into *Science And Spirit,* the first version of this book. While retaining much of the earlier material, the current work reflects the understanding I have reached during the ensuing years. In retrospect, the memory of the images remains unchanged, however, my understanding of the meaning has matured with the following years to contemplate, ponder, and meditate upon the conclusions.

In the current work, the apodicticity of the argument extends much further into the realm of consciousness, awareness, and spirituality than I initially imagined, proving, in effect, that if you

[3]The word 'apodictic' comes from the field of formal logic. It refers to an assertion that is incontrovertibly or demonstrably true as a result of having been deduced systematically, in a stepwise manner, from incontestable foundational principles, in such a way that denying it leads to a contradiction or an absurdity.

believe in computer microcircuits and the atom bomb, you are *forced* to admit to the reality of an eternal spirit. Indeed, pretentious though it might seem, I do not think it unreasonable to claim that this book forces us into an indisputable understanding of the meaning of life.

Let me put it differently: If you believe in science, then you need not fear death, in any form. If you wish to truly *believe* this statement, read on.

Introduction

Most people are pretty certain they have an intuitive prehension that there is more to life than the mundane and material, and term that unquantifiable *something* spirit. Whether they even admit to the existence of a transcendent soul in the orthodox sense, even the most stolid atheist is willing to admit that the human mind may represent something that seems to exceed the sum of its material components. Without quantification, however, our grasp on the spirit and the spiritual remains tenuous, for its ultimate foundation remains that slippery notion of faith.

All mystical and spiritual disciplines concur that unshakable conviction is the most important key to success in their individual endeavors. All of the rituals, incantations, and paraphernalia of the various practices are, in most cases, merely props to bolster the practitioner's belief system to the point where mind can rally its full force without the obstacle of doubt. To more primitive societies, these rituals may indeed have served to impart the force of conviction, but our modern world-view entails entirely different impediments to belief.

Faith came more easily to our forbearers. Most modern people, educated in the doctrines and certainties of science, look toward spirituality, the paranormal, and the supernatural with a kind of wistful wariness. The yearning to embrace some manner of spirituality cannot be dispelled, but we fear that doing so amounts to intellectual suicide. The result is a spiritual void in our lives, leaving unanswered the issues of what we may hope for, how we should live, and the ways in which we may control the circumstances of our existence. The primary purpose of this book is to help patch that hole by showing that we need not choose between archaic doctrine and modern certainty. Hence, this book is an attempt to build a logical, apodictic argument that will lead the reader to the point from which such enlightenment may be directly assimilated.

The problem we face is that there is a piece missing from our understanding that precludes a truly apodictic level of certainty when it comes to all matters spiritual, mystical, or paranormal in nature. One purpose of this book is to provide that missing link so that these precepts become a matter of absolute certainty, for only thus can the mind be freed to actually experience the true nature of being.

Our goal, then, is to remove the roadblock to understanding represented by modern scientific knowledge (or, to be more precise, the shallow, reductionist approach to science propounded by many of its less visionary practitioners). This is accomplished by showing how modern scientific principles are fully consistent with both Western and Eastern spiritual concepts, as well as most mystical and paranormal phenomena. With the support of scientific certainty, the reader is freed from intellectual anxiety and uncertainty, and is thus better able to open up to true inspiration and unreserved belief.

Faith has come to have a rather negative connotation in the world of science. It is often interpreted as belief in an idea despite overwhelming evidence to the contrary. As such, it is commonly associated with issues like creationism versus evolution, or assertions that the earth has existed for a mere few thousand years. This is unfortunate, and not at all reasonable. All of our greatest minds achieved their distinction by 'thinking outside of the box'. What is this, but faith—faith in themselves; faith in ideas that, at the time, were not supported by the prevailing attitudes and knowledge of the day?

Indeed, is not the alternative to be like the historically arrogant Lord Kelvin (1824-1907), whose hubris is best represented by his 1900 address to an assemblage of physicists at the British Association for the Advancement of Science, in which he stated that: "There is nothing new to be discovered in physics now. All that remains is more and more precise measurement." Even earlier examples that typify the monumental pomposity of this individual are provided by his 1895 assertion that: "heavier-than-air flying machines are impossible," followed by his 1896 statement: "I have not the smallest molecule of faith in aerial navigation other than ballooning...."

Although somewhat more circumspect about their pronouncements, most members of the scientific mainstream share a similarly smug attitude. Although they admit that they do not have *all* of the answers, inwardly they are highly confident that they have all of the important ones, and that new discoveries will undoubtedly consist of greater and greater detail.

Fortunately, a few visionary individuals (such as Stephan Hawking, John Wheeler, J.S.Bell, David Bohm, and Hugh Everett III) have argued most convincingly that the experimentally proven characteristics of reality not only permit, but in most cases, demand the assumption of mechanisms that could result in what we currently label as spiritual, mystical, or paranormal phenomena.

One of the first impediments we will need to transcend is that slippery item known as 'intuition'. We all think we have a pretty good intuitive understanding of what constitutes reality, but, as we shall see, that understanding has been proven time and time again to be utterly unreliable.

After all, what is this thing we call 'intuition', but the collected prejudice of what our sensory experience has revealed over the years of our conscious lives. Thus, to most of our ancestors, the flatness of the earth was intuitively obvious, as was the 'fact' that the sun moved around the earth.

When we look around us, we perceive a three-dimensional reality, populated by objects that reflect a small cross-section of the electromagnetic spectrum we call 'visible light' only because, in evolutionary terms, our predators and our prey perceived and reflected that certain narrow band of the electromagnetic spectrum, existed within the 4-dimensional space-time continuum, and so forth. There was simply no survival benefit in experiencing the world around us in any other way. But does it represent anything resembling 'ultimate reality'? Judging by all of the implications of modern science, the answer is 'no'. What we sense as reality, therefore, is no more or less valid than, say, the multiple images seen by a common housefly. Hence, intuition is really nothing more than the sum total of all our experiences from a severely prejudiced viewpoint.

Einstein's theories of Special- and General Relativity, Quantum Theory, Bell's Theorem, Hugh Everett's Many-World Model, String Theory, Chaos Theory, and many other disciplines require us to accept a very different world view than that suggested by our limited sensory apparatus. So, in order to provide a firm foundation for the arguments to follow, let's begin by examining some of the most prevalent theories of modern physics in greater detail.

PART I: THE RELEVANT SCIENCE

The mental image of physicists as sterile, plodding, anti-religious, non-spiritual beings; or alternatively, as mad scientists bent on world destruction, is one that must be overcome. The true nature of physics is best conveyed by remembering that the original name for this discipline was 'Natural Philosophy'. This label fits even better now than when it was more commonly used. Although the issues discussed in this section may, at times, seem only marginally related to God, spirit, and religion, they collectively form a firm foundation for the explanations and explorations that follow. I make no attempt to present the entire depth and breadth of each subject, but rather, focus on those aspects that contribute to a convincing picture of the subject matter of this book—the nature of spirituality and the meaning of life. By the end of this section, the reader will have gained an appreciation for the fact that the world around us is very different from what it appears. With this knowledge will come a correspondingly expanded vision of reality.

Mathematics and Reality

Although I won't be taxing the reader with much in the way of actual mathematics, most of the concepts we'll explore in the coming chapters derive primarily from mathematical foundations. Therefore, it is essential that that we share a common understanding of what mathematics is.

Mathematics, in its most profound sense, is more than a mere computational tool. Rather, it is a concise, consistent, and reliable means for thinking about that which would otherwise defy contemplation. It is a method that imposes discipline, but not restriction, on our thought process. It constrains our cogitation to fruitful directions, not by coercion, but by quickly showing us potential dead ends, without the necessity of extended excursions and the subsequent retracing of steps. Far from restricting the imaginative process, it opens up an infinity of otherwise inconceivable vistas.

To most of us, math falls into one of the following categories:

1. The course(s) you took in high school or college, barely passed, and for which you never had any subsequent use,

2. A tool that you use in your job as an engineer, actuary, architect, etc., which is about as mystical as a hammer.

To others, such as physicists and mathematicians, mathematics holds a more central position in their daily lives.

Be that as it may, what can be said with assurance is that mathematics is at the core of the subjects we are about to discuss, and as applied, provides a level of confidence that is far more rigorous than simple philosophizing. It can sometimes be difficult for the non-mathematician to appreciate the level of certainty inherent in such proofs, or alternatively, just how certain we can be of our *uncertainty*, when such is the case. Thus, for the non-mathematician, a degree of faith is required that when, for example, Newton stated his three laws of motion, behind this deceptively simple verbalization exists a prodigious and convincing body of mathematical support.

Throughout this text, although we will be discussing many concepts that are inherently mathematical in nature, the presentations will avoid the actual mathematics to the greatest possible extent. There is, however, one topic that *will* require just a bit of mathematical development in order to support subsequent topics, so let's get that out of the way now. I encourage the reader not to be put off by what may be a few unfamiliar figures (just think of them as magical runes, if you like), and to make an effort to slog through this section. I promise that it's actually more conceptual than computational, and the effort will be well rewarded. This essential topic is something known as *imaginary numbers.*

For those unfamiliar with them, imaginary numbers (also known as complex numbers) are those which have the form:

$$a + bi,$$

where i is $\sqrt{-1}$.

Recall from introductory algebra how the square root of 1, written $\sqrt{1}$, is equal to ± 1; that is, multiplying (-1) times (-1) or (+1) times (+1) both equal 1. Thus, there is no *natural* number that, when multiplied by itself, equals −1. Consequently, we need another way to designate that mathematical entity. This entity is simply referred to as i. We can illustrate this graphically as shown in Figure 1.

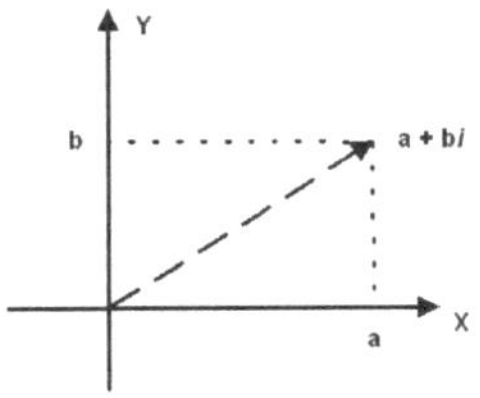

Figure 1

The plane in which resides all complex numbers is called the complex plane, or Gauss-plane. In this illustration, the points corresponding to the real (non-imaginary) parts of the number (e.g., 'a') are located on the X-axis. The points corresponding to the pure imaginary parts of the number (e.g., 'b') are located on the Y-axis. Therefore we say that the X-axis is the real axis and the Y-axis is the *imaginary* axis. The important thing to understand here is that the imaginary axis is not at all 'imaginary' in the ordinary sense of the word. There are myriad applications of so-called 'imaginary' numbers in the physical sciences and engineering disciplines.

It's also important to understand that the imaginary plane does not represent a spatial dimension, such as height or width or depth. If we were to create a three-dimensional graph, showing height, width, and depth, the imaginary axis would have to be in the fourth dimension. Since there's no way to even envision such a thing, it's easier to plot the imaginary dimension relative to a single real axis. (Similarly, when we look at String Theory's ten-dimensional symmetry, it's to be understood that the complex plane is orthogonal to all ten.)

Complex numbers are essential for many areas of mathematics and engineering. Of particular interest to the present discussion is the fact that the central formulation of quantum mechanics, which describes the behavior of matter at its most basic level, describes a wave with a complex, or 'imaginary' component! Most mainstream physicists initially believed that this component had no physical analogy, but today, many of the cutting-edge interpretations rely heavily on a physical assumption of imaginary space. We'll return to this concept when we examine the notion of the astral body, and possible models for PSI interactions.

Classical Mechanics

Classical Mechanics represents the world of common sense and intuition. It is the realm of solid objects that behave in predictable ways when acted upon by definable forces. The central statement of classical mechanics is a system of equations known as Newton's Laws of Motion. Newton summed up the laws of motion by showing that:

1. A moving object will continue in uniform motion until something acts to divert it. (Although this may seem perfectly obvious to us, at the time it was as counterintuitive as the most abstruse notions of space-time curvature. After all, observations of the real world, and hence, intuition, tell us that every moving object eventually slows down and stops. This was what the ancient Greeks believed because, being ignorant of the concept of friction, it was how their senses told them things really worked.)
2. Any change in the motion of a moving object depends on a force acting on the object, and is opposed by the mass of the object. (This, too, is far less obvious than it may appear to the modern mind. Remember Galileo's experiment in which he dropped two objects, one heavy and one light, from the Leaning Tower of Pisa? At the time, conventional wisdom dictated that the heavier object would *obviously* hit the ground first. Galileo demonstrated that this was not the case, but it took Newton to explain why: although gravity acts on an object with a force

proportional to its mass, the object resists with precisely this same inertia.)

3. For every action there is an equal and opposite reaction. (Here again, it took Newton's formulation of the mathematics of calculus to rigorously demonstrate why this is so. Although a person diving off of a small boat would certainly have noticed that the boat recoils or tips as a result of the diver's departure, it would have taken a visionary mind indeed to speculate that the earth itself recoils, though infinitesimally, as well.)

Although, stated verbally, these laws sound deceptively simple, it's important to understand that in Newton's development they entail a considerable degree of mathematical rigor. In fact, Newton had to develop an entirely new branch of mathematics, known as differential calculus, just to support his derivation of these laws.

This classical theory of physics, which reigned supreme from the time of Newton until the twentieth century, provided an orderly model of a world made up of objects moving and pushing each other around in predictable ways. Although the mathematics could be difficult, the basic ideas meshed with our common-sense notions of how the world works.

One implication of classical mechanics that caused considerable consternation, particularly among the church leaders of the time, was determinism. Basically, classical mechanics said that if one could identify the starting conditions and all relevant forces in a given system, then theoretically, one could determine every detail of that system at any time in the future. Thus, by implication, it appeared as if every aspect of the future would have been mechanically predetermined at the moment of creation. If this were the case, then free will would be totally illusory. Humans were seen as no more than aggregates of particles—a complex system, to be sure, but a mechanical one nonetheless. In this view, humans were no more capable of self-determination than the mechanism inside a pocket watch. This determinism formed an intrinsic part of the prevalent scientific paradigm of the nineteenth century.

Starting at the beginning of the twentieth century, however, emerging physical theories began to include concepts that defied common sense. Despite their counterintuitive nature, these theories could not be easily dismissed because they consistently made accurate predictions about phenomena that could not be explained with Newtonian physics. Gradually, and despite considerable resistance, physicists were forced to accept the resulting bizarre new world-view.

As we shall see, the Newtonian view was largely superceded by the implications of the so-called 'modern physics', particularly Quantum Theory. Nevertheless, it may well be that the deterministic notions of Newton are closer to the truth than most present-day physicists imagine. When we examine Hugh Everett's multi-world interpretation of quantum mechanics, we'll see how all possibilities may indeed be pre-ordained and deterministic. Free will, however emerges as we navigate experientially through the myriad of possible worlds. More importantly, we will also find that Quantum Theory—hailed as the "most successful theory in the history of science," *demands* that we accept that human volition underlies all physical phenomena!

One final point before moving on: When we say that Special Relativity or Quantum Theory has 'replaced' Newton's classical mechanics, it should be understood that in the mathematical formulations, these original laws of motion emerge when dealing with macroscopic systems, or objects whose relative velocity is slow compared with the speed of light. In Einstein's own words:

"Creating a new theory is not like destroying an old barn and erecting a skyscraper in its place. It is rather like climbing a mountain, gaining new and wider views, discovering unexpected connections between our starting point and its rich environment. But the point from which we started out still exists and can be seen, although it appears smaller and forms a tiny part of our broad view gained by the mastery of the obstacles on our adventurous way up."[1]

[1]Albert Einstein and Leopold Infeld. *The Evolution of Physics*, New York, Simon and Schuster, 1938, p152.

Some Field Theory

The first major expansion of Classical Mechanics was the concept of the force field, ushered in by James Clerk Maxwell. Maxwell began by studying magnetic and electrical phenomena, from which he deduced the basic formulation of all electromagnetic waves. In Maxwell's formulation, the field replaces the simple force of Newtonian mechanics. Maxwell's equations provided part of the foundation for Einstein's subsequent development of his theories of Relativity.

Mankind had known about magnetism since the time of the ancient Greeks, who were familiar with a particular kind of rock, called magnetite, which possessed the property that if two chunks of it were brought together in certain orientations, the rocks would stick together. In August, 1831, Michael Faraday discovered that passing a magnet through a loop of wire created a current. This discovery, known as electromagnetic induction, changed the world by demonstrating how electrical energy could be created from mechanical motion.

Around 1865, while attempting to explain the mechanisms of induction, Maxwell deduced his theory of electromagnetism. Basically, what Maxwell found was that a changing electric field has the effect of inducing a magnetic field. This, in turn, opposes the electric field. The magnetic field then induces an electric field of the opposite polarity, which in turn damps the magnetic field. The net effect is a wave of alternating magnetic and electric fields, which propagate at the velocity of light. In technical terms, the resulting equations described all electromagnetic phenomena as consisting of orthogonal, mutually inductive, electric and magnetic fields, as shown in Figure 2.

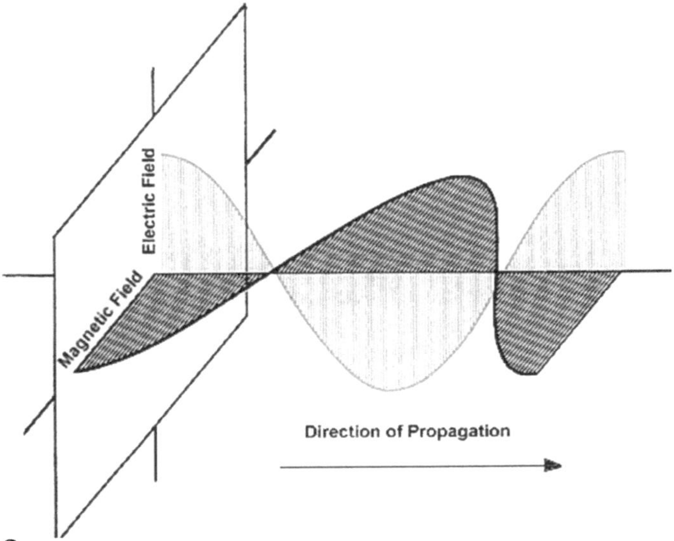

Figure 2

This of course is an oversimplification. To really create a convincing description would entail many pages of elaboration that would certainly bore the most indulgent reader. However, for the purpose of demonstrating how concise the language of mathematics can be, I'll go back on my word and throw a few more equations at you:

$$\nabla \bullet \mathrm{E} = \rho$$

$$\nabla \bullet \mathrm{B} = 0$$

$$\nabla \mathrm{X} \mathrm{E} + \frac{\partial \mathrm{B}}{\partial t} = 0$$

$$\nabla \mathrm{X} \mathrm{B} - \frac{\partial \mathrm{E}}{\partial t} = j$$

Think of this along the lines of attending the performance of an opera in a language you don't understand. You can still appreciate the experience without following the particulars of the dialog. Alternatively, you can look at these equations as figures engraved on a talisman, which convey, to the initiated, some profound secret of the universe, which, indeed, these particular runes do.

What's significant in this case is that these four statements alone completely describe the entire mystery of electromagnetic phenomena. Through a bit of mathematical manipulation, these equations also predict the speed of light. As it turned out, the value thus derived was identical with the measured speed of light. This led Maxwell to conclude that light, magnetism, and electricity were all simply different aspects of the same underlying force described by his equations. (Again, this may seem obvious to us now, but in the mid-1800s, it was a staggering leap of imagination.)

The importance of this development was that it formed one of the starting points from which Einstein deduced his theory of Special Relativity, and *that* has some very important implications for understanding spiritual and paranormal phenomena. Moreover, through some rather hoary calculations, it is also possible to derive *Schrödinger's Equation*, one of the cornerstones of Quantum Theory, from these same four statements. Such 'coincidences' strongly suggest we are dealing with something that really does reflect an intrinsic aspect of reality.

Once Maxwell had established that light behaves like a wave, it was widely assumed that it must propagate *through* some medium. This medium was known at the time as the 'luminiferous aether'. This led to one of the most famous experimental failures in history.

In 1887, Albert A. Michelson and Edward W. Morley attempted to measure the velocity of the earth through this luminiferous aether. They used two beams of light, split from a single source and reflected at right angles. This would allow very small changes in the speed to be seen as an interference pattern when the beams were recombined. (Interference is the effect produced by the combination or superposition of two systems of waves, in which these waves reinforce, neutralize, or in other ways interact with each other. We'll be looking more closely at interference patterns shortly.)

Michelson and Morley's idea was that the speed of the light

moving in the direction of the earth's motion would be retarded by the aether, and this would manifest itself as a misalignment of the recombined beams. Their reasoning went something like this: If the earth is moving through this substance, light should logically be slowed down in the direction of travel, much as a boat's actual speed is less when traveling upstream against a current. Their failure to detect such an effect was considered a paradox, and ultimately lead Albert Einstein to conclude that:

a) there is no such thing as the aether, and,

b) the speed of light is constant under all circumstances.

These assumptions formed a cornerstone of Special Relativity, which will be our next topic of discussion.

Special Relativity

As most modern readers are aware, Einstein's theory of Special Relativity is associated with the equation $E=mc^2$. But what exactly does this mean? Actually, the theory, and even the mathematics (which we won't be going into), are comparatively simple; it's their ramifications that are difficult to fathom.

Relativity actually has more in common with classical mechanics than those branches generically referred to as 'modern physics', such as quantum mechanics. This is because relativity is *deterministic*. Complex as it may appear, it is a philosophy through which we can theoretically know everything there is to know about the system to which it is applied.

In the previous chapter we discussed how James Clerk Maxwell developed the field equations for electromagnetism, of which visible light is one example. For the current discussion, the most significant outcome of this effort was the revelation that the speed of light is equal to approximately 186,000 miles per second. As mentioned earlier, Michelson and Morley attempted to detect variations in the speed of light that should logically have resulted from the earth's movement through the luminiferous aether, which was, at the time, assumed to permeate all of space.

Einstein's bold departure was to take a subtle change in orientation by recasting this paradox as an assumption--he started out by *assuming* that the speed of light simply is a constant for all observers. Likewise, he assumed that nothing could move faster than light. This is important: the absolute velocity of light was *an assumption*, and moreover, was one predicated on a whole slew of other implicit assumptions, including an assumption of three spatial dimensions and one time dimension. Indeed, many modern cosmological theories hold that entire dimensions may exist in which time plays no part whatsoever. Since velocity is defined as the change in position over *time*, it is easy to see that Relativity can not be a universal law of nature, merely one that appears to apply in our illusory 4-dimensional space-time continuum. (Indeed, scientists now know that there was a juncture at which the universe itself was expanding faster than the speed of light.)

The second element that went into the formulation of Special Relativity was the classical concept of relativity, originated by Galileo some two hundred years earlier, and subsequently quantified by Lorentz. Basically, these formulations state that the laws of mechanics are valid for all frames of reference in uniform motion.

For example:

If you drop a ball, it falls straight down, then bounces straight back up. You can repeat the experiment while standing on the open flatcar of a train moving at sixty miles per hour, and from *your* perspective, the ball still drops straight down and bounces back up. From the perspective of a person standing at the station, however, the ball appears to describe a 'V' shape, as it moves downward and forward (in the direction of the train's motion), and then upward and forward. Nevertheless, knowing the speed of the train, one can easily explain the difference between the two perspectives using a fairly straightforward and intuitive mathematical formula (known as the Lorenz Transformation).[1]

Everyday experience with modern highway travel provides a wealth of intuitive examples. If you're cruising down the road in your BMW at ninety miles per hour, and up ahead you see Grandpa

[1]Basically, what Einstein did was to take an existing, well-understood, and fairly elementary equation, add an assumption, and work through the ramifications mathematically. The math is within the grasp of anyone who's completed elementary calculus. This is not to in any way diminish Einstein's insight, but rather, to demonstrate that it was not the result of incomprehensible mathematicizing, but rather, to insightful reasoning and logic. The mathematics merely confirm the validity of the result. The result is that if you make these assumptions, and you work through the math, *then* E-MC^2, which subsequently was proven true by the atomic bomb.

Yokem in his pickup truck doing fifty, it's easy to see how you appear to be approaching him at forty miles per hour (90-50). Similarly, if Carl Crotchrocket is traveling in the other direction on his motorcycle at a hundred and ten, he will appear to be approaching you at a speed of two hundred miles per hour (110+90). The point here is that as long as we know the uniform speeds and directions, we can perform the simple math and come up with reasonable, common-sense answers. We start running into problems, however, when light enters the picture.

Let's suppose that instead of driving at ninety miles per hour, you're actually traveling ninety miles per second (maybe you've installed the latest aftermarket supercharger). Carl Crotchrocket's new superbike is traveling at one-hundred and ten miles per second in the opposite direction. Suppose Carl turns on his motorcycle's headlight as you're approaching each other. The classical transformation says that if you could measure the speed of light as you drive, the result you should get would be 186,000 + (110 + 90), or 186,200 miles per second, right?

Of course, nearly everyone now knows that this is wrong. We would still measure the speed of light as 186,000 miles per second , and therein lies the seed from which special relativity bloomed under Einstein's careful gardening.

By assuming the truth of this statement, (that is, the constant velocity of light) the mathematics led Einstein to several results that were curious to say the least:

The faster you go, the heavier you get.

The faster you go, the shorter you become, in the direction of your motion.

Mass and energy are equivalent (e.g., a hot iron is heavier than a cold one).

The faster you move, the slower your clock will run.

We needn't go into the math that proves these effects. All we need to know is that, 'common sense' notwithstanding, they have been repeatedly and satisfactorily proven in the years since Einstein postulated them. For example, in 1972, using the world's most accurate atomic clocks, scientists definitively proved a measurable difference between the flow of time for a clock circling the earth in a jet plane and an identical, synchronized clock 'at rest' on the earth. Thus, Einstein showed that there is no such thing as a universal, absolute time.

Pause here and reflect that every time a scientist applies relativistic time corrections at an accelerator lab, she or he is demonstrating a perspective or viewpoint *outside* of time. We need not, indeed, can not envision such a thing other than in the most abstract fashion, or through the aid of our mathematics, but this does not invalidate the perception; indeed, it *confirms* the consistency of our reasoning.

By far, one of the most profound outcomes of Special Relativity is that, although it *seems* as though we have control over our movement in the three spatial dimensions, while time flows inexorably by, this is not what happens at all. We experience events unfolding with the passage of time, but this, according to Albert Einstein, Stephen Hawking, Louis de Broglie, and virtually all modern physicists, is merely an illusion. If we could view the four-dimensional continuum in its entirety, we would see that past, present, and future 'already' exist in their totality, embedded, so to speak, in the fabric of space-time. Although we *appear* to discover new slices of space-time as successive aspects of the material world, the ensemble of events that constitute the space-time continuum exist prior to our awareness of them. This is a crucial observation, and a perspective that is central to the thesis of this book. We will be repeating it in a variety of ways because, as you will see, it leads to some very significant conclusions with respect to paranormal possibilities and spiritual awareness. Hard to envision as it may be, experimental evidence thus far seems to indicate that this perspective is essentially correct.

Although this world-view would appear to argue for a level of determinism even more onerous than the mechanistic assertions of classical physics, such is not the case, as we shall see shortly.

These are pretty heady concepts, but they are essential for an understanding of the true nature of our world. There are many aspects of modern physics that seem to violate our intuition, so it's important to bear in mind that intuition cannot be relied upon in the face of proven observations to the contrary. Our task, then, is to stimulate our instinct so that *these* realities become an integral component of our intuitive repertoire.

Special Relativity predicts other unusual effects that can also seem highly counterintuitive. One of our main difficulties is that to fully appreciate what's happening, we have to imagine looking in on three-dimensional space from a fourth dimension that we can never quite envision. To make the descriptions a bit easier to understand, it's useful to imagine ourselves looking in on a two-dimensional world from our more familiar three-dimensional perspective (see Figure 3).

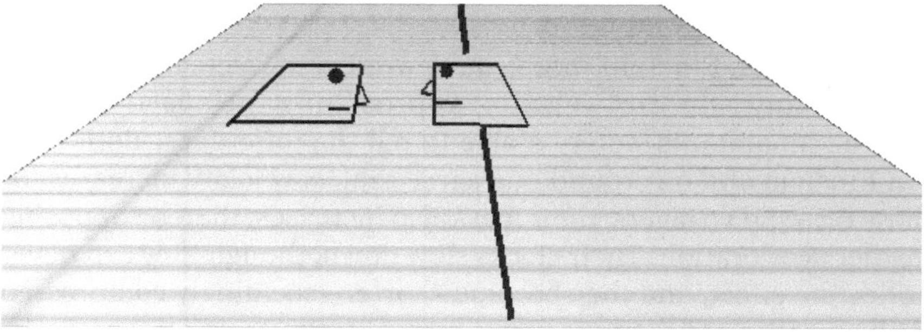

Figure 3

Here we see Mr. Flat and Mr. Shallow, two denizens of a two-dimensional world (Flatland) that resembles a sheet of paper. In the picture, Mr. Shallow is standing in a doorway, and Mr. Flat is unable to get through without rudely pushing Mr. Shallow out of the way. From our more privileged perspective, we can easily see that Mr. Flat might pass by going above or below the sheet of paper. Unfortunately, the inhabitants of Flatland can never even imagine such a possibility, just as we can't envision moving through a fourth spatial dimension.

(Lest this appear to be no more than a wild flight of fancy, I point out that it's a common problem assigned in university physics classes. One can devise all laws of mechanics, electrodynamics, and so on for a two-dimensional universe, and all are completely consistent.)

Now, let's suppose that this entire flat world is moving very fast. Let us imagine, also, that our three-dimensional space is full of thick jelly, and that suspended in this jelly are tiny red-hot rivets. As we move the sheet through the jelly, every time the sheet encounters a rivet, it burns a tiny, almost imperceptible hole. The flat people never notice an individual hit, however, the cumulative effect is obvious: things break down and eventually are no more. Flat people, also, degenerate and die. Everything, in fact, tends to go from the very orderly—the nice flat sheet of paper—to the chaotic, at which point the sheet finally looks like a slice of Swiss cheese. This is something familiar to the flat people, and to their parents, and to all of their ancestors, back to the first flat amoebae. It's so familiar that they even have a name for it: they call it entropy, and use it as their primary measurement of the apparent passage of time. This, of course, is illusory, but it has the advantage of appealing to a necessarily inexact intuition.

As flat-Einstein might have predicted, we can move this world at any arbitrary velocity, approaching the speed of light, and the inhabitants won't be able to tell the difference. Both their flat Einstein and our round one agree that as velocity increases for these intrepid voyagers, their mass also increases, thereby opposing further acceleration. Their clocks slow down, though they can never notice the effect, and they become shorter—seemingly compressed in the direction of their motion.

This last effect, according to Einstein, is an illusion. What appears as a compression, is actually a rotation. To understand this better, hold a flat sheet of paper vertically in front of your face. Then slowly rotate it until only the edge is visible. From our vantage point in three dimensions, it's easy to see what's really happening, but from the viewpoint of one of the Flatlanders, it appears as if the space voyager has been squashed. Notice, also, that since less of the rotated paper encounters the hot 'rivets of time,' it would also appear to age more slowly, that is, its clock slows down.

The problem with these images is that they still try to view time from a perspective within time itself. Once we step outside of time, our broader perspective notices that time itself is a rather static entity. The entirety of time, from its beginning in the 'big bang' to its ending within black holes, and at the supposed final 'big collapse', is an *instantaneous*, or rather, an *untimeable* phenomena. It is only the motion of our sequential awareness through the static fixture of space-time that causes the illusion of flowing time to emerge.

Though harder to envision than flatland, we can imagine a similar situation in which our three-dimensional world interacts with a continuum of higher dimensionality. To understand the mechanisms involved in the prohibition against faster-than-light travel, imagine dropping a marble through a deep vat of viscous liquid (older readers may recall a television commercial in which a pearl was dropped through a bottle of shampoo). The falling object gains speed, up to a certain point at which the force of gravity is precisely counteracted by the frictional resistance of the liquid. The marble reaches what is known as its *terminal velocity*, after which it can fall no faster. We may think of the four-dimensional space-time continuum as having a sort of 'viscisity' that opposes movement beyond the light speed. The important thing to bear in mind is that this restriction only applies *within* that continuum. Scientists now believe that there may be entire universes in which time plays absolutely no part whatsoever!

We'll be examining these issues more closely a bit later, and will also be revisiting our friends the Flatlanders. For now, there's one additional point to be made about Special Relativity. Sacrilegious as it may seem, some of Einstein's assumptions may not have been totally justified. In particular, he made a tacit assumption that we exist in a universe consisting of just three spatial dimensions and one time dimension. This is not to say that Einstein was *wrong*. We just need to bear in mind that the results and predictions of his theories will apply only to the extent that his assumptions remain in effect. In Newton's case, his laws of motion are still perfectly correct, just so long as the objects we're dealing with move slowly with respect to the speed of light (and are large with respect to the size of an energy quanta). Similarly, Einstein's Relativity yields perfectly usable results, just so long as we limit the field of application to the three spatial dimensions, plus time, with which we are normally concerned.

Later, when we examine the 'many-world' interpretation of quantum mechanics, and Bell's theorem, we will be compelled to reconsider these assumptions.

Heisenberg and Uncertainty

When we examined classical mechanics, we saw that one of the key aspects of Newton's laws was that, given sufficient information, we could always determine where an object is, how fast it is moving, and predict where it will be at some time in the future. This approach worked fine for things like cannon balls, or even lunar landing modules, and was perfectly adequate for several hundred years. The picture changed dramatically, however, when scientists started examining the behavior of very small 'objects', such as electrons.

The first person to notice and quantify this difference was Werner Heisenberg. What Heisenberg discovered, in essence, is that we can never know the precise position and speed (or more properly, the momentum) of a particle, because in measuring one, we change the other. The first important thing we need to understand about this uncertainty, which is known as the Hiesenberg Uncertainty Principle, is that it describes an intrinsic aspect of nature, not some limitation imposed by our experimental procedures or instrumentation. As a result, the more certain we are about one measurement, the less we can know about the other.

Let's look at a simple arithmetic example:

X * Y = 100

X	Y	X * Y
1	100	100
2	50	100
4	25	100
5	20	100
10	10	100
20	5	100
25	4	100
50	2	100
100	1	100

Figure 4

As we can see in Figure 4, if the value of X increases, the value of Y must decrease in order to ensure that the product is always equal to 100. In its mathematical formulation, the Heisenberg Uncertainty Principle takes precisely this form when describing the relationship between position and momentum.

The second critical aspect of the uncertainty principle is that it is directly related to the size of the 'packets' or quanta that describe the basic substance of all matter and energy. To be exact, the uncertainty principle states that:

dp*dx > $h/2\pi$

The critical component in this expression is the '*h*', which is known as Plank's Constant, and describes the size of an energy quanta. All energy exchanges anywhere in the (our) universe must occur in integral multiples of *h*. Thus, Plank's Constant describes the basic granularity of reality.

In the above expression, 'dp' and 'dx' designate the change of momentum and position, respectively, over time, which is to say, our uncertainty of either parameter. The fact that everything on the right side of the expression (everything to the right of the '>' symbol) is a constant, means that the product of the position and the momentum components must always be greater than that constant. Thus, as in our previous arithmetic example, if one of these items gets bigger, the other must get smaller, right? So, for example, if we measure x (the particle's position) exactly, then dx (the uncertainty in the particle's position) approaches zero, and hence the uncertainty in momentum (dp) must approach infinity in order to maintain the value of their product. Conversely, if we measure precisely the particle's momentum, our uncertainty as to where it is becomes infinite.

The result of this uncertainty is that, in the quantum mechanical world, we can never predict where a particle will be and what it is doing with 100% certainty. We can only speak in terms of probabilities. For example, we can say that an atom will be at some location with a 99% probability, but there will be a 1% probability it will be somewhere else. (In fact, there is a small but finite probability that it will be found across the universe.)

A consequence of this quantum mechanical nature of the world is that particles can appear in places where they have no right to be (from an ordinary, common sense [classical] point of view). For a real-world example of this phenomena, consider the so-called quantum-well problem:

Let's begin by examining how objects behave in the world of classical mechanics.

An object can store energy as the result of its position. For example, the heavy ram of a pile driver stores energy when held at an elevated position. This stored energy of position is referred to as gravitational potential energy. In general, gravitational potential energy is the energy stored in an object as the result of its vertical position (i.e., height). This energy is stored due to the gravitational attraction between the Earth and the object. The gravitational potential energy of the heavy ram in a pile driver is dependent on two variables—the mass of the ram and the height to which it is raised. There is a direct correlation between gravitational potential energy and the mass of an object: more massive objects have greater gravitational potential energy. There is also a direct relation between gravitational potential energy and the height of an object: the higher an object is elevated, the greater the gravitational potential energy. These relationships are expressed by the following equation:

$E_p^{grav} = \text{mass} * G * \text{height}$,

where G is a constant representing the gravitational attraction of the earth, and E_p^{grav} is the gravitational potential energy.

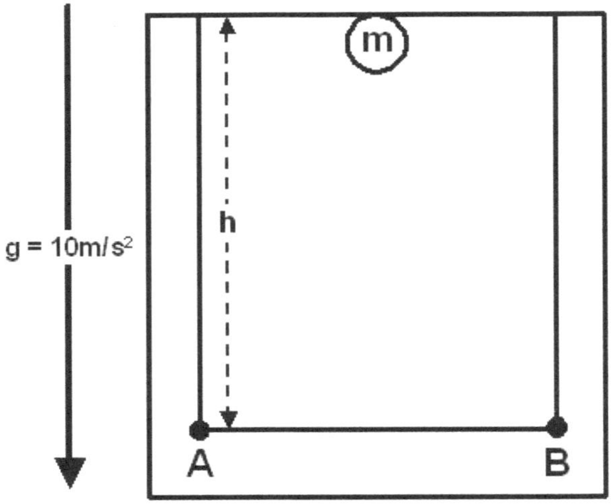

Figure 5

Suppose we have a well, as shown in Figure 5. We begin by holding a ball so that its upper surface is precisely aligned with the top edge of the well. Using classical mechanics, the ball has a certain amount of potential energy, given by the mass of the ball, times the height, times the gravitational constant, as previously described.

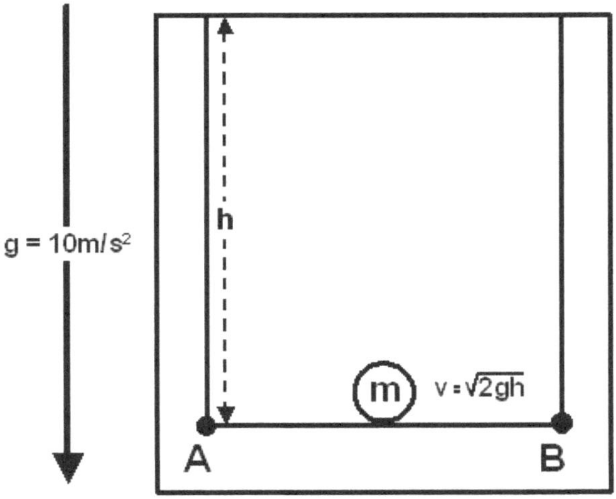

Figure 6

If we release the ball, it will fall into the well, gaining velocity as it goes. When it reaches the bottom (Figure 6), it will (for our purposes) have no more potential energy, however, it will then have kinetic energy as a result of its velocity. This will be given by the formula:

$$E_K = (1/2)mv^2,$$

where v is the velocity the ball has attained by the time it reaches the bottom of the well. Since no energy was added (i.e., we didn't *throw* the ball downward), this will be equal to the original potential energy. We can figure out the velocity if we know the original height (h)—it works out to:

$$\sqrt{2gh}$$

(As an aside, note that the mass, ***m*** has dropped out of the equation. This is why Galileo's famous experiment, in which he dropped two different masses from the Leaning Tower of Pisa, proved that two different weights would fall at the same rate.)

Ignoring friction, the ball will then bounce, reversing direction, and rise back to its initial position, as shown in Figure 5. So long as no additional energy is added to the system, the ball can never get out of the well. This is because rising above the height of h would require more energy than it started with. Thus, using classical mechanics, the probability of the ball ever finding itself beyond the area delineated by points A and B is precisely zero.

We can, in various ways, set up a similar situation using an electron instead of a ball. In this case (known as the quantum-well problem), however, the result is quite different. Because the laws of classical physics don't apply, we have to use the rules of quantum mechanics instead. Quantum Theory says there is a definite non-zero probability our electron will find itself outside of the well. In fact, its position could be anywhere out to infinity!

This happens to have been one of the first problems I had to deal with right out of college, working at Digital Equipment Corporation's Advanced Semiconductor Development Lab in Hudson, Massachusetts. We were developing the DEC Alpha chip (then the fastest microprocessor in the world), and as it turns out, this exact phenomenon started showing up in the ultra thin-film transistors being used in the high-speed circuits. As you can easily imagine, a misplaced electron translates into a misplaced bit, which can wreak havoc in the world of digital logic. What we discovered after many hours of reviewing our basic architecture was that the inexplicable errors were, in fact, due to that pesky electron tunneling phenomenon. So, the answer is: "yes, these effects are definitely real." This, however, is by no means the strangest phenomenon predicted by Quantum Theory!

Waves and Particles

As is the case for each of the chapters in this section, the subject of Waves and Particles could easily fill an entire textbook. From this plethora of possibilities, the information presented will focus on those areas that demonstrate what the issue is, why you should believe it, and how this information serves to support the model of a spirit-enabled universe.

The concept of a particle is pretty intuitive, and corresponds roughly to the kinds of objects we discussed earlier, in the chapter dedicated to Classical Mechanics. Basically, a particle is a very small *thing* that exists at a discreet point in space and time.

Waves are also pretty easy to envision. We've all seen waves in water, such as those on the ocean, or the ripples that result from tossing a pebble into a pond. A vibrating guitar string represents a special kind of wave, in which the end nodes are fixed. This is known as a *standing wave*. Waves appear as regular patterns of energy that are distributed through space. This distinguishes them from particles, which exist at a single point in space. Other kinds of waves include sound waves, which propagate as alternate compressions and rarefactions of air pressure, and light waves, which exist as alternating electric and magnetic fields.

One interesting thing we can do with light waves is to shine a light through a narrow slit cut in a piece of paper. If we place a screen on the other side of the slit, the result will be a fuzzy band of light that is brightest in the center, directly behind the slit, and gradually fades away, as shown in Figure 7.

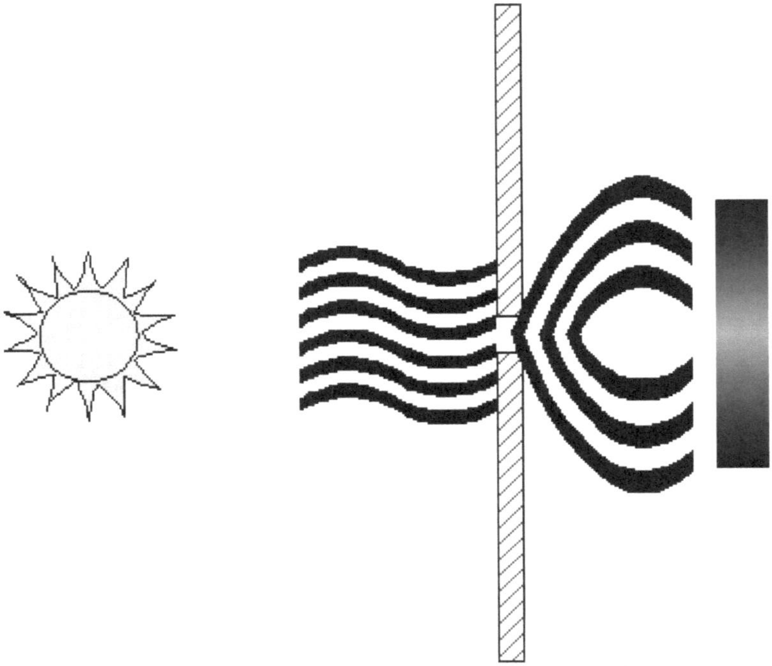

Figure 7

Nothing too startling here, right?

Okay, suppose instead of one slit, we have two slits. The result is a series of light and dark bands, as shown in Figure 8.

Figure 8

This effect isn't quite as intuitive as the single-slit situation, but it's still pretty easy to understand. Basically, what you end up with is bands of light where the waves reinforce each other, separated by darkness, where the waves cancel each other out. This is known as an interference pattern. You can see a similar effect by tossing two pebbles into a pond at the same time. Musicians are familiar with the audio equivalent, which occurs when two notes, played simultaneously, are slightly out of tune from each other, the result being a series of audio pulses or beats.

Now, here's where things start getting interesting.

In 1905, Albert Einstein wrote a paper on photoelectric effect (which won him the Nobel Prize in 1921). In this paper he demonstrated that light actually consists of a stream of *particles* called *photons*. Conceptually, what Einstein did was shine a light on a metal surface, and measure the electrons that were produced (in the form of a current). Einstein reasoned that if light really behaved like a wave, the time lag between shining the light and measuring the current should increase as the light intensity is reduced. This would happen if, like pumping on a playground swing, it took a certain period of time for the light wave to 'pump' energy into the electrons until they were finally separated from their associated atoms. What Einstein discovered, however, was that this does not happen. What he did find was that the time lag remains constant, but the number of emitted electrons decreases. Thus, Einstein concluded, what actually occurs is that as the light is dimmed, fewer photons are produced, and each photon acts like a tiny bullet, which knocks a single electron out of its orbit. (Subsequently, there have been thousands of other experiments that prove this is actually what happens.)

That should have elicited a great big "Huh?" After all, how can something be both a wave, which is distributed throughout a region of space, and also a particle, which exists at a single point in space? The answer, which really doesn't make the issue any more intuitive, is that it depends on how you look at it. If your experimental apparatus is set up to examine wavelike phenomena, like the one shown in Figure 8, light looks like a wave. If, on the other hand, you examine photoelectric effect, then light looks like a particle.[1] This

[1]Shortly thereafter, Louis de Broglie demonstrated that electrons, in addition to behaving like particles, are also wave-like in nature. The phenomena of electrons behaving as waves is responsible for the operation of the electron microscope (and many other applications).

suggests that neither conceptualization is actually real, and both actually reflect one underlying reality.

Einstein, of course, was aware that the phenomena of diffraction and interference can be explained only on the basis of the wave picture.[2] He never even attempted to dispute the fact that these two interpretations of the nature of light were completely contradictory. In fact, he simply took it on faith that the apparent contradiction would ultimately be understood.

Now, here comes an even bigger 'HUH?':

Suppose you set up your experimental apparatus as shown in Figure 8, but use a light source that can spit out a single photon at a time, or use electrons instead. Rather than a screen, we'll use a photographic plate to record our results. After running the experiment with a large number of photons, we find that we end up with the exact same pattern of dark and light bands as when we passed a steady light through the two slits. Next, we cover up one of the slits, and repeat the experiment. The result this time is a photographic plate that looks more like the background shown in Figure 7.

So?

Well, this means that at least some of the photons fell on the area of the plate that would have been totally black (no photons) if both slits had been left uncovered.

So?

Well, this means that, somehow, these photons must 'know' that the other slit has been covered!

HUH?

[2]Diffraction occurs when, for example, light is passed through a prism, and is separated into its constituent wavelengths or colors. Interference, as explained earlier, occurs when two waves join together, and their crests and troughs either reinforce each other, or cancel each other out.

There is, in fact, no definitive answer to this question, which represents one of the central paradoxes in modern physics. When we come to the subject of nonlocality, we will discover that instantaneous (that is, faster than light) communication between distant particles is a necessary outcome of Quantum Theory. Special Relativity, remember, insists that nothing, not even information, can propagate faster than light. Quantum, on the other hand, insists that information *must*, under certain circumstances, travel instantaneously. This strongly suggests that, as stated before, Relativity must apply only within a limited region of a much larger reality. As we shall discover later, the properties of this greater reality are precisely those needed for supporting spiritual, mystical, and paranormal phenomena.

As disturbing as some of these concepts may at first appear, aren't they closer to what life is really like? Our problem is that we've worked hard to convince ourselves that idealized particles with well-determined properties really exist. We've come to believe that the simplified assumptions made by Newton are a reflection of reality, rather than useful tools for getting approximate results. Sure, they're close enough for hitting our enemies with cannon balls, and even for sending probes to Pluto, but do they tell us anything about reality? Obviously not! Experience tells us that life is not comprised of absolutes. Things are never black or white, but always various shades of gray.

A useful analogy might be to compare the wave-particle duality to the notion of conic sections. When we take a familiar conic section, and intersect it with a plane, the result may be a circle, as shown in Figure 9.

Figure 9

Alternatively, it may also appear to be an ellipse, as shown in Figure 10.

Figure 10

Then again, as we can see in Figure 11, the intersection may also represent a parabola.

Figure 11

Finally, as Figure 12 shows, the intersection may describe a hyperbola.

Figure 12

As these examples demonstrate, even in elementary three-dimensional geometry, what we see very much depends on our perspective and orientation. In the more complex quantum world, what we perceive as waves or particles (or space-time curvature) may similarly depend on our orientation. In this case, however, perspective depends on how our local three-dimensional region of space-time intersects a higher-order dimensionality. Thus, while we may seem to see different things, there *is* an underlying reality—we simply can't see it directly, at least not by relying on our five senses.

Getting back to our clairvoyant photon paradox, we can repeat the double-slit experiment using a stream of electrons, and we get the same kind of results, because all particles can also manifest themselves as waves.

One of the 'answers' to this problem, as mentioned earlier, is a concept known as nonlocality. (Einstein disliked this notion, and referred to it as "spooky action at a distance.") Nonlocality means that, from some perspective we cannot understand, there is a connection between the photon (or electron) and the experimental apparatus. The difficulty with nonlocality is that the 'message' is transmitted instantaneously, and if this is so, then information is being transmitted faster than light. This, of course, is prohibited by Einstein's theory of Special Relativity, so it's not too surprising that he found the concept difficult to accept.

Another more mainstream (though no more intuitively satisfying) interpretation, is that the particle actually goes through both slots at once, and only by looking do we force it to manifest one state or the other! (In the parlance of quantum mechanics, we say that observation forces the particle's *wave function* to collapse.) We shall delve into this interpretation more deeply in the next chapter. This is a central tenet of quantum reality: until the system is actually observed, the particle exists only as probability of manifestation--something called the wave function.

The wave function is simply an equation that expresses the probability that a particle will be in one state or another *when observed.* (Remember when we were discussing imaginary mathematics, and mentioned that the wave function has a complex, or imaginary component? In a sense, we can think of the quantum wave as something 'vibrating' between real and imaginary space. When the wave function collapses, its imaginary component disappears; that is, we have forced the 'entity' into manifesting itself as a concrete object in three-dimensional space. We'll have more to say about this later. For now, we may simply surmise that the timeless complex plane may represent the route through which these subtle connections propagate.)

At the moment of observation, the wave function is said to collapse into one of its possible states, thereby revealing the corresponding object. Quantum mechanics insists that nothing can be said about the particle until it is actually observed.[3] This, in effect, makes it impossible to separate the observer from the experiment. It should also begin to suggest that mainstream science describes a world in which, theoretically at least, intent and will could have a direct influence on subjective reality.

It must be emphasized that a central doctrine of Quantum Theory is that this statistical formulation of the laws of physics does not reflect ignorance of the physical state of affairs. The situation, scientists insist, is intrinsically different than those in which statistics are employed by actuaries or gamblers as a means of distilling large volumes of inherently knowable data. Rather, the statistical aspect of Quantum Theory represents (in the orthodox interpretation) a fundamental limitation that governs reality at the most intimate level. This view (the Copenhagen interpretation) is held by most of the mainstream scientific community, and states, in effect, that nature is inherently probabilistic; that is, that there is nothing else that *can* be known.

It may well be, however, that our use of statistical probabilities in dealing with physical processes does indeed reflect a kind of ignorance. This ignorance may result from our inability to comprehend the basic nature and structure of the universe, perpetuated by a propensity for clinging to topologically simplistic, three-dimensional assumptions about the structure of reality. Although the distinction may seem moot, what we interpret as a probability that a certain particle will manifest itself at a specific time or place may actually represent a probability that we will be looking in the right place at the right time. In this sense, it is a likelihood related to the observer, not that which is being observed. This is more consistent with what Special Relativity tells us about time; that is, the particle is 'already' manifest, we just happen to be looking there at precisely that moment.

[3]This is actually a statement of The *Copenhagen Interpretation*, which is the most conservative statement of Quantum Theory.

Are these theoretical musings real? Well, so far as we can speak of anything being 'real', we can say that they are consistent with observable phenomena.

In 1935 Erwin Schroedinger, one of the pioneers of quantum mechanics, proposed one of the most celebrated paradoxes in Quantum Theory to illustrate this concept. This paradox came to be known as *Schroedinger's Cat*.

Schrödinger's Cat

Schrödinger's Cat is a way of illustrating one of the central paradoxes of quantum mechanics. It purportedly shows that when we observe something, we also physically change it, but most modern views are far stranger still.

Schrödinger proposed a thought-experiment in which a small amount of radioactive substance is placed into a container. The quantity is precisely measured so there is a fifty-percent probably that it will decay, emitting an alpha particle, within the period of one hour. A Geiger counter is set up to detect this emission, should it occur.

Nearby, we place a live cat in a box, and with it, a device that will release a fatal dose of deadly hydrocyanic acid if the Geiger counter registers the radioactive emission. (Calm down, all you animal-rights people—it's only an *imaginary* experiment, okay?) Thus, logic would suggest that after one hour the cat will be alive if the radioactive element has not decayed, and dead if it has; that is, there's a 50/50 chance the cat will *either* be alive *or* dead. This is the view dictated by classical logic, however, quantum mechanics demands a far different interpretation.

Schrödinger claimed that the moment before we look inside the box, the system possesses a wave function that describes the cat as existing in some mixed state of dead and alive. Essentially, Quantum Theory demands that the wave function for the emitted particle, and hence, the state of the cat, does not assume one of its two final conditions until it is actually observed. Thus, the cat is, somehow, both alive and dead until the split second when the experimenter opens the box, looks inside, and finds either a live or a deceased feline. When and where does this half-alive half-dead cat change its state from quantum zombie into one definitive state? Who collapses the wave function? The cat, the Geiger counter, or the physicist?

Consider Frank R. Stockton's story, "The Lady Or The Tiger?"

A young commoner who dares to fall in love with the king's daughter is subjected to a perplexing punishment. He is compelled to open one of two doors. Behind one is the princess he has dared to love. Behind the other, a ferocious tiger. If he selects the door with the princess, they will be immediately wed. If he chooses otherwise, he gets torn to shreds.

As bad as our poor suitor's situation appears, the quantum problem is even worse. In the fable, the suitor knows that a lady is behind one door, and a killer tiger is behind the other. He may not know which door leads to which, but the answer is inherently knowable. In the situation of Schrödinger's cat, no knowledge of the system inside the box will ever let you predict the cat's fate. Whether it's alive or dead is inherently unknowable until the box is opened.

If this situation is reminiscent of the Heisenberg Uncertainty Principle, give yourself an A+, because that's exactly what's happening in the Schrödinger's Cat thought experiment. By checking to see if the cat is alive, we actually cause a physical change in the cat. What this means is that there's a difference between a 100% dead cat, a 100% live cat, and a feline that is 50% alive and 50% dead. Quantum reality, therefore, possesses its own kind of physical being, halfway between possibility and actuality. The 'things' that comprise the essential building blocks of matter are not, as we might imagine, independently existing entities, but rather, a set of relationships that reach outward to other things.

The point Schrödinger was trying to make by proposing the cat paradox was that we cannot say anything regarding 'the cat itself' before looking. Our knowledge of the cat after observation is the only reality we can ever experience, or so he believed at the time. The proposition that Quantum Theory represents a complete theory, even though it can't tell us anything at all about what is actually happening, is a restatement of the Copenhagen interpretation mentioned in the previous chapter.

Can a cat really be both dead and alive?

In the January 2000 issue of the prestigious *British Journal of Nature*, physicists David Wineland and Chris Monroe reported successfully creating an experiment similar to that described by Schrödinger's Cat. They used a beryllium atom in a 1/2mm trap, and then excited it with laser beams in a precisely controlled way. As a result, they were able to create a situation in which the atom could be shown to be in both quantum states at once for 10^{-4} (one one-thousandth) second. During that period (which is a relatively long time in subatomic terms), the experimenters were able to demonstrate that the two instances of the single atom were interfering with each other. If the particle had been exclusively in one state or the other, no such interference pattern would have been possible.

Scientists hailed the Wineland-Monroe experiment as a crucial step toward building a quantum computer. A quantum computer would make today's fastest supercomputers seem like counting on our fingers by comparison. According to Wineland, a quantum computer of just 300 quantum bits (atoms) could contain more information than there are particles in the whole universe.

Need more proof?

In August 2000, IBM announced that it had developed the world's first advanced quantum computer, which used 5 atoms as its processor and memory, thereby demonstrating the potential of this new computer technology (and the reality of the underlying principle). In so doing, IBM and Wineland-Monroe have effectively proven that the cat is, indeed, both dead and alive. As these advances demonstrate, the paradoxical worldview proposed by Quantum Theory is more than just an academic's pipe dream.

An explanation proposed by physicist John Wheeler is that 'reality' may not be wholly physical. In some sense, our cosmos might be a participatory reality, requiring the act of observation (and thus consciousness itself) in order to be fully instantiated. If this is true, we create not only the observation, but even the reality itself.[1]

Sound too far out to be real?

[1]It should be noted here that we refer to apparent reality as experienced in our four-dimensional space-time continuum. To the extent that these are reflections of a hypothetical greater reality, that *true* reality would remain unchangeable.

Well, to demonstrate his conjecture, Wheeler proposed a *delayed-choice* version of the two-slit experiment. Basically, this would entail waiting until *after* the particle has passed the two slits, and *then* deciding whether to cover one slit or not, thereby creating a particle pattern or wave-interference pattern. This would imply that an action performed in the present could influence something that has already happened in the past. Stated differently, the electrons would seem to know in advance how the experimenter would choose to observe them later.

Another egghead's daydream? Well, the experiment was actually carried out in mid-1980s by two groups working independently of one another, and both confirmed the theory. (Bear in mind that the forefront of scientific discovery are almost never visible to the general public. As startling as these discoveries may be, we can safely assume that there are far more astounding experiments being conducted in secret government laboratories.)

Although these experiments serve to confirm that *something* strange is going on in the quantum world, they do little to explain *how* it is happening.

One of the first to consider how these bizarre phenomena might occur was Doctor Hugh Everett, III. In 1957, Everett, at that time a Princeton University graduate student of the well-known American physicist John Wheeler (mentioned above), proposed the theory now widely known as 'the many-world interpretation of quantum mechanics'. (He preferred to refer to it as his 'relative state' formulation or 'Theory of the Universal Wavefunction'.) This theory may seem even more bizarre than those just discussed, but in fact, has much to recommend it. Everett proposed that time branches off into every possibility, each just as real as the single one that we happen to (think we) experience. His doctoral thesis on the subject has been considered a valid scientific theory ever since its publication. Although only a minority of physicists initially considered it, it has yet to be disproved, and is in fact becoming more widely accepted.

Everett considered it unreasonable that a seemingly 'magical' process, such as the collapse of the wave function, should be used to explain an aspect of nature, while at other times, systems follow normal continuous laws. In Everett's interpretation, neither collapsing wave functions nor observer knowledge are needed to explain the experimental results of quantum mechanics. Instead, at every occasion where a quantum event has more than one outcome, the universe *seems* to split.[2] Stated differently, the wave function is fully instantiated in all of its myriad of possibilities—a *Universal Wave Function*. We have one universe in which the cat is alive, and another where it is dead. The effect of our observation is to force us to *experience* one branch or the other (or at least that's how it seems to us from within).

From the multi-world viewpoint, the universe is like a tree that splits and re-splits into a multitude of new sub-trees with every quantum event. Each of these new universes has a slightly different quantum 'history'. Because an observer happens to have followed one particular path through the diverging branches of this universe-tree, he never perceives the splitting. Instead, he interprets the resolution of the various possibilities into one particular outcome in terms of the classical wave-function collapse. The observer, however, plays no active role in the splitting.

Formally, this model is developed with a considerable degree of rigor using a branch of mathematics known as Hilbert Spaces. (A Hilbert Space is a nonlinear vector space[3] that can have an infinite number of dimensions. The dimensions of the Hilbert Space have nothing to do with physical dimensions as we commonly understand them, but mathematically represent the collection of all analytic functions.) The mathematics are quite unambiguous: the other Everett-worlds *do* exist, and furthermore, they occupy the same space and time as we do.

The resulting mathematical description of the universe is complex, but some physicists—Dr. Stephen Hawking among them—have successfully validated, and even expanded upon this model. In Hawking's interpretation, the universe is completely self-contained. It is a closed and finite container without boundary or singularities; that is, there is no 'edge' to the space-time continuum. (We will examine this type of topology in more detail later.) The result of Hawking's model is that space-time is finite, but unbounded

[2]When we consider what we have already discovered about time, it's easy to see that "splitting" infers change over time, so the apparent forking paths are in fact an intrinsic aspect of the static structure of space-time.

[3]Mathematically, a linear *vector space* is a nonempty collection of algebraic objects, which can be manipulated in accordance with certain well defined procedures, specifically, addition and multiplication. A *nonlinear* space requires only that the constituent points possess the property that for any two points, there be an associated scalar value that represents the distance between them.

(the boundary condition of the universe is that it has no boundary). The universe is neither created nor destroyed. It simply *is*.

When you stop to consider what Special Relativity says about time, this new interpretation makes perfect sense. If time is created at the moment of the Big Bang, and if we can imagine a perspective from which we can 'see' the past, present, and future, then the 'beginning' and the 'end' of the universe are simply illusions, forced upon us by our viewpoint from 'inside' of time.

The upshot of the Everett-Hawking model is that everything that logically can happen, does happen, no matter how improbable. Once again, the reason we aren't aware of these alternative paths is that once we seem to choose one, the others appear inaccessible.

According to our present knowledge of physics, although it is possible to detect the presence of other nearby worlds (i.e., through the existence of interference effects, as previously described), it is impossible to communicate with them. Mathematically, this corresponds to a well-proven property of all quantum theories known as linearity. Linearity implies that the worlds can interfere with each other from the perspective of an external observer, but the interfering worlds can't influence one another. (An experimenter in one of the worlds can't arrange to communicate with his own, already split-off, quantum copy in another world, at least outside of esoteric means.)

Non-communication between the splitting Everett-worlds also explains why we can never be aware of the splitting process itself, since such awareness would imply communication between worlds. As we shall see, with just a small conceptual modification, we can devise a convincing theory for a universe in which paranormal concepts fit rather intrinsically. First, however, we need to take a quick look at Bell's Theorem.

Bell's Theorem

Albert Einstein was never comfortable with quantum mechanics precisely because of the strange features we've just discussed. To his mind, these theories described a reality inconsistent with his intuition of how a proper universe should operate. In 1935 Einstein, together with Boris Podolsky and Nathan Rosen, published a list of objections to quantum mechanics that has come to be known as *The EPR Paper*. High on their list of complaints was nonlocality, or what Einstein referred to as "spooky action at a distance." Nonlocality means that interactions not possible through simple contact are somehow enforced, faster-than-light, across space and time. Naturally, since Einstein's theory of Special Relativity began with the explicit assumption that *nothing* could exceed the speed of light, nonlocality was a pill he found difficult to swallow.

The EPR paper, through a flawless chain of logical induction, showed how, under certain circumstances, quantum mechanics predicts that performing an action on one particle can instantaneously influence another particle arbitrarily far away. The intent of this exercise was to demonstrate that, since the conclusion was *obviously* absurd, Quantum Theory was inherently flawed. The resulting uproar in the physics literature went on for decades as the founders of quantum mechanics scrambled to come to grips with the EPR criticisms.

Almost thirty years later, Irish mathematical physicist J.S. Bell proved that the results predicted by quantum mechanics were inherently inconsistent with any theory that preserved locality. In other words, if you set up an experiment like that described by Einstein, Podolsky, and Rosen, and you get the results predicted by quantum mechanics, then there is no way that locality can be true. In 1972, John Clauser actually performed an experimental version of Bell's theory at the University of California. Though his intent had been to disprove Bell's Theorem, much to his chagrin, the results actually confirmed Bell's predictions. The experiment was subsequently repeated in 1982 by Alain Aspect at the University of Paris, with the same results.

In the years since its development, Bell's Theorem has been called the most important discovery in the history of science. The implications of this theorem, and the experimental findings that have repeatedly confirmed it, are staggering. They force us to conclude that the entire notion of a purely objective world is in conflict not only with the theory of quantum mechanics, but also with experimental evidence. These findings point insistently to a profound interaction between conscious mental activity and the physical world itself. Among the implications of Bell's Theorem is confirmation of the many-world interpretation of quantum mechanics. (Other noteworthy scientists endorsing the many-world interpretation include Stephen Hawking and Nobel Laureates Murray Gell-Mann and Richard Feynman.)

So, what does the universe really 'look like' at this point? From Einstein's Special Relativity, we've seen that time and space form a single continuum, in which past, present, and future are instantaneously existent. From Quantum Theory, and in particular, Everett's interpretation thereof, we understand that this continuum may actually consist of every possible variation of the past, present, and future. This is actually somewhat comforting, since it saves us from the bleak philosophy of determinism. From Bell's contribution, we have not only a reaffirmation of Everett's interpretation, but also, proof that the space-time continuum itself is interconnected in ways that are not immediately obvious from our normal three-dimensional perspectives.

Before moving on it's important for us take a step back and understand that as bizarre as these conclusions may sound, they follow a pretty solid mathematical trail that, thus far, nobody's been able to disprove. Even by the most conservative scientific interpretations, what's beginning to emerge is a picture of a universe that is definitely far more complex than our poor old intuitions have led us to believe. Does the resulting picture begin to suggest a universe in which spiritual, mystical, and paranormal phenomena could be integral components? As we shall see, the answer is yes.

General Relativity

When we discussed Einstein's theory of Special Relativity, you'll recall that the venerable scientist started with two assumptions. The first was that the speed of light represents a constant upper limit for the velocity anything can attain. The second was that the laws of physics must be the same for observers in any two frames of reference moving at a constant velocity with respect to each other. Basically, what this second assumption meant was that Carl Crotchrocket wasn't allowed to jam on the brakes, speed up suddenly, or make any turns. (In other words, acceleration, which is defined as any change in velocity over time, was not permitted.)

Despite these limitations, a whole bunch of interesting stuff emerged, such as space-time, atom bombs, clocks that run at different rates, and so forth. Nevertheless, Einstein wasn't satisfied with this limitation on acceleration. In other words, he wanted to generalize the theory of relativity. The result was something known as—you guessed it: *General Relativity*.

As before, Einstein steadfastly (obstinately, perhaps) maintained that nothing could travel faster than light. This time, however, he assumed that the laws of physics should be the same for any two frames of reference, whether or not one was accelerating with respect to the other. When considering these issues, Einstein was struck by what he later described as the happiest thought of his life; namely, that an observer who is falling from the roof of a house experiences no gravitational field. This led him to conclude that a gravitational field was completely equivalent to a corresponding accelerated reference frame (that is, when you step on the accelerator pedal of your car and feel yourself pressed against the back of the seat, the experience is indistinguishable from lying on your back and being pulled down by gravity). Once again, following these simple ideas through to their logical conclusion led to some rather startling predictions, and a whole new way of looking at the universe.

In essence, what Einstein deduced was that the fabric of space-time is not flat and level like a plane, but rather, is radically curved and covered with bumps and gullies. Without going into the gruesome mathematics, or even the details of development, the upshot of all this is that, according to General Relativity, there's actually no such thing as gravity. Rather, what we *perceive* as gravity is really a result of space-time being bent by the proximity of mass. (Actually, this is not entirely accurate. What Einstein preferred to believe was that there was no such thing as mass, and that what *seems* like mass is only an artifact of these space-time 'dimples'.)

Although this may seem like a rather abstruse notion, it really is a lot more reasonable than trying to imagine how a mass (e.g., the sun) exerts an instantaneous affect on another mass (the earth, for example) through this nebulous gravitational force field that nobody has been able to explain. General Relativity compels us to conclude that there is actually no reality (at least with respect to our local 4-dimensional continuum) except space-time curvature. Remember, mass and energy are interchangeable (Special Relativity), and particles are intrinsically the same as waves (Quantum Theory), and since mass is a result of space-time curvature . . . well, that's all there is folks.

Let's run through that once again, just to be sure we've got it:

Space-time curvature results in what manifests itself as mass, according to General Relativity. Mass and energy, Special Relativity tells us, are interchangeable aspects of the same thing. From Quantum Theory, both mass and energy are expressed by the wave function as tendencies to exist. These waves manifest themselves as particles when they interact in a very small space, over very short periods of time. The aggregates of these particles are perceived by our three-dimensional sensory apparatus as material objects—but this is all an illusion.

Sounds like a pretty far-fetched fairy story, doesn't it? Well, that's all it would be if Einstein hadn't envisioned several ways to prove it, and darned if experiments haven't repeatedly shown that this actually seems to be the case. Fortunately, these proofs are pretty easy to understand, so let's take a look at them and see how convincing they really are.

The question that immediately came to mind the first time I read about the implications of General Relativity was: "How in the world could you tell the difference?" After all, if you spend your whole life living in a curved universe, you'd never be able to notice it; or would you? Well, the fact is, you would.

Since it's extremely difficult for us to get our minds 'outside' of three-dimensional space, it will be easier to understand if we once again imagine what it would be like to live in a world of two dimensions, rather than three.

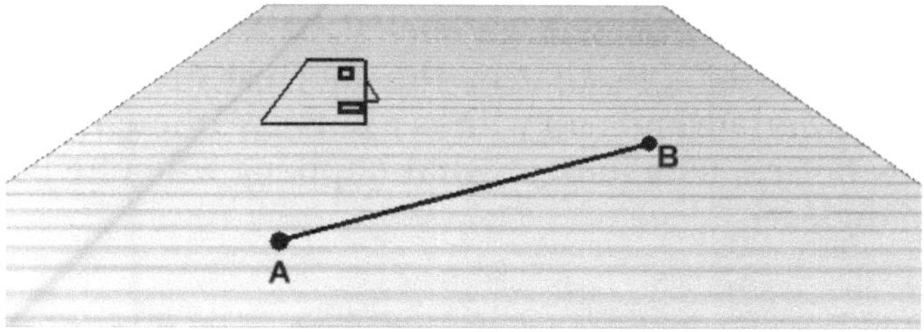

Figure 13

Figure 13 shows a denizen of Flatland—our hypothetical world that exists on a sheet of paper. Mister Flat may dream about what might exist above or below the surface of his two-dimensional world, but he can never directly experience it. Similarly, we can imagine dimensions beyond our familiar three, but can't even picture them (that's why we're using the Flatland for our example).

The Flatland people can develop a complete system of geometry, just like ours, and the rules will be exactly the same. For example, a straight line represents the shortest distance between two points, as shown in Figure 13. Similarly, when a flat mathematician progresses to the study of triangles, he will discover that the three internal angles of every triangle always add up to precisely 180 degrees.

Now, let's play a dirty trick on our flat mathematician. While he's asleep, we'll pluck him off of his nice flat sheet of paper and deposit him on the surface of a sphere, as shown in Figure 14.

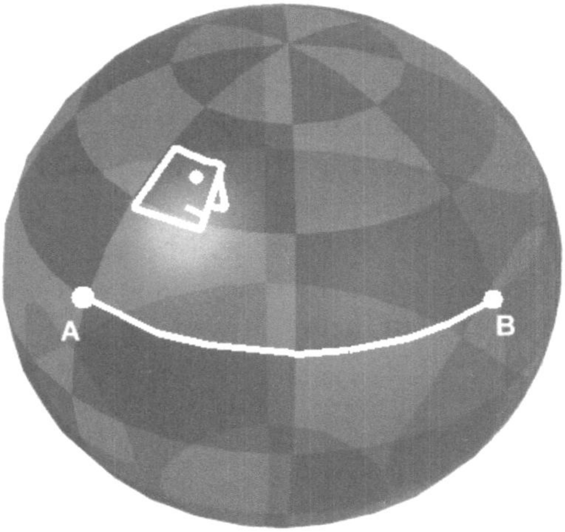

Figure 14

Now when our flat geometry expert draws what he thinks is a straight line, that is, the shortest distance between two points, what he ends up with is a curve. Although this might escape his notice, let's look at what happens when he begins to study triangles:

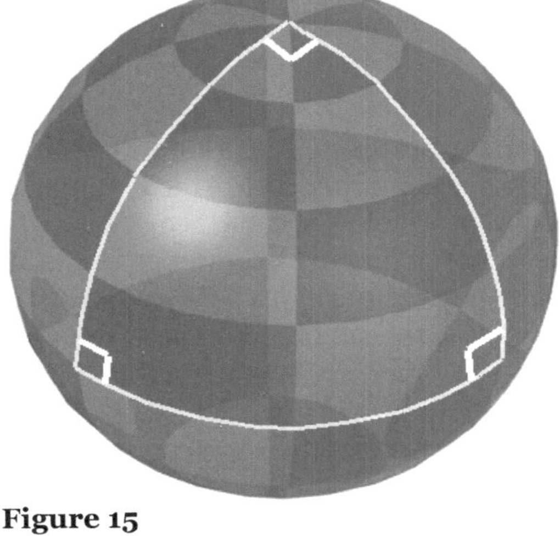

Figure 15

What he'll find is that the sum of the interior angles of his triangles is always greater than 180 degrees. This may be difficult to notice in small figures, but as they get larger, the effect will become increasingly obvious. In the extreme case, as shown in Figure 15, he'll find he has described a triangle with three right angles, or a total of 270 degrees!

Now, this is an effect that *we* are accustomed to. We live in three-dimensional space and recognize the fact that we live on a planet that is roughly spherical in shape. The Flatlander, however, still believes he's living on a flat sheet of paper. His initial inclination might be to theorize the existence of some mysterious force that stretches his lines in every direction, but he will never be able to get a handle on precisely how this nebulous force operates. Eventually, a flat Einstein may come along and recognize that he's not living in a flat Euclidean plane, but rather, in one that is intrinsically curved.

Now, back to our own three-dimensional world.

One of the first things Einstein predicted was that light should appear to bend in the region of the sun. He suggested that astronomers might take photographs of an easily recognized star cluster (in the night sky, when they wouldn't be influenced by the sun), and note the positions of each star relative to the others. Then, during a time of total eclipse, they could photograph the same cluster, and see if the stars nearest the sun seem to be in the wrong position. As it turned out, there was a convenient solar eclipse just four years after Einstein formulated his theory, and guess what? Two separate researchers found that light was bent precisely as Einstein had predicted. (Since then, the effect has been confirmed many times, using ever more sophisticated instruments.)

The second phenomena that Einstein predicted was what we've come to call a 'black hole'. A black hole is thought to be the remnant of a giant star (approximately three times the size of our sun, or greater). According to the theory, when the sun's fuel is finally exhausted, it will collapse in on itself, forming a body so dense that nothing, not even light, will be able to escape. Although nobody has actually seen a black hole directly (in fact, since no light can escape, it couldn't be seen anyway), astronomers are pretty sure they've discovered such stellar bodies by studying the wobble of certain stars in distant galaxies. When two stars are located close to each other, the mutual gravitational effect causes them to rotate around their combined center of mass. What the astronomers look for is a single star that behaves as if it's part of a binary star pair, but which has no visible counterpart.

The third prediction of General Relativity is referred to as gravitational red-shift. The theory predicts that regions of high gravity will slow down time. One of the results of this is that light waves (which, in a sense, are like tiny clocks, because their frequencies are time-dependent) should show a measurable shift toward the red end of the spectrum in the vicinity of high-gravity objects. This, too, has been verified experimentally.

Finally, although Einstein didn't actually predict this one, it was found that General Relativity accounted for anomalies in the orbit of Mercury, which until that time had defied all explanations that depended on classical physics.

Thus, although we can never really know for sure, all evidence seems to indicate that General Relativity accurately models the topology of our space-time continuum.

So, to sum up, classical mechanics claimed that the earth revolved around the sun because the earth, like any mass, tends to move in a straight line unless acted upon by an external force. That force is the gravitational attraction of the sun, which keeps pulling the earth toward it, much like a yo-yo being spun around on a string. What General Relativity tells us is that the earth revolves around the sun because that's the path of least resistance in the curved space-time created by the masses of sun and earth. A useful analogy would be walking around the base of a steep mountain, rather than climbing up one side and down the other.

Although none of what we've discussed thus far can be considered incontestable fact (in the final analysis, nothing can be stated with absolute certainty), it has been proven with a high degree of confidence. Beyond this, many scientists believe that space-time may be even more radically curved than merely a few bumps. We'll look at some of these possibilities later. (I separated these discussions because I wanted to make sure we had a clear distinction between widely accepted theory and more radical conjecture.)

String Theory

As suggested earlier, the relationship between Relativity and Quantum Theory was strained from the start. As things stand today, scientists recognize three fundamental forces in the universe:

Gravitation—the force that appears to control the interaction of objects based on their mass,

The strong nuclear force—which is responsible for holding particles together at very close ranges (this is why protons in the nucleus of atoms stick together, even though their like charges tend to push them apart), and

The electroweak force—a unification of electromagnetism and the so-called weak nuclear force.

In practice, General Relativity is used for calculations involving gravitation, and quantum mechanics is used for everything else. To the orderly mind of a physicist, this is a highly unsatisfactory state of affairs. What scientists would *like* to see is one set of tools they can use in every circumstance.

As in the previous chapter, General Relativity describes gravity and mass as effects of space-time distortion. General Relativity is a mathematically elegant solution, that appears to have passed all experimental tests. Nevertheless, its inability to dovetail with Quantum Theory is considered a fundamental flaw. This is because the preponderance of experimental evidence accumulated over the past century strongly suggests that nature obeys the tenets of Quantum Theory. As we have seen, the salient features of Quantum Theory are that:

1. Energy may only exist in discrete quantities (a physicist would say that energy is quantized), and
2. There are inherent uncertainties in the prediction and measurement of physical systems.

General relativity differs fundamentally from Quantum Theory in both of these areas. In General Relativity, energy is *not* quantized, and even more importantly, there is no inherent uncertainty associated with either the prediction or the measurement of objects in space-time. Hence, General Relativity and Quantum Theory are mutually incompatible.

Under most scientific situations, this really isn't much of a practical problem. When dealing with the very small scales at which Quantum Theory is applicable, gravitational forces are so small as to be negligible. Similarly, at macroscopic scales, the probabilities inherent in Quantum Theory become essentially deterministic. Thus, the incompatibility would not be a practical problem, but for one pesky phenomenon: the black hole.

Under conditions that occur inside black holes (or at the instant of the *Big Bang*), the influence of gravity attains a scale that is comparable to that of the electroweak and the strong nuclear forces. Thus, in order to understand the origin of the universe, the quantum nature of gravity must be taken into account in a manner that is consistent with Quantum Theory.

Unlike General Relativity, Quantum Theory does not recognize space as anything but flat. Quantum theory is based on elementary particles (objects with no structure and no dimension), interacting with each other. Despite their seemingly ethereal nature, these elementary particles do possess certain inherent properties that distinguish one from another, and result in their characteristic interactions.

In order to bring these two theories into harmony, physicists have tried (without success) to recast General Relativity in terms of Quantum Theory by attempting to represent gravity as a force carried by a hypothetical particle called a *graviton*. Unfortunately, these attempts have led to nonsensical results. Physicists refer to this as *the divergence problem of quantum gravity*. Without delving into the gristly mathematical details, the cause of the divergence problem can be traced to Quantum Theory's assumption of particles as single points.

String theory attempts to resolve this incongruity by describing all matter as consisting of vibrating one-dimensional objects with length and tension. (Since these one-dimensional objects don't have length, as we commonly envision it, the notion of a string should not be taken too literally.) Furthermore, string theory seems to promise to unite the other (electroweak and strong) interactions as well, paving the way toward the Holy Grail of theoretical physics: the so-called *Grand Unification Theory*.

Among the objections typically leveled against string theory is the fact that it requires physical dimensions in excess of our familiar 4-dimensional space-time continuum. In fact, it requires, at a minimum, a ten-dimensional continuum consisting of nine spatial dimensions and one time dimension. This might help clear up the apparent contradiction represented by a string of only one dimension. How? Let's go back to our friends the Flatlanders. Suppose we were to take a sharp pin and gently touch the point to the flat world. To our two-dimensional observers, it would appear to be a one-dimensional point, although *we* could see that it possesses an extent in three-dimensional space that they cannot directly perceive.

In terms of our own experiential three-dimensional universe, scientists have speculated that the six extra dimensions required by String Theory are actually folded into three-dimensional space at scales too small for experiments to detect. (This is called *compactification*.)

Yet another hint at the existence of higher-order physical dimensionality comes from the study of *quasicrystals*.

In classical crystallography a crystal is defined as a three-dimensional periodic arrangement of atoms with translational periodicity along its three principal axes. Thus it is possible to obtain an infinitely extended crystal structure by aligning building blocks called *unit-cells* until the space is filled up. In 1984, however, Shechtman, Blech, Gratias & Cahn published a paper that marked the discovery of quasicrystals. In it, they showed electron diffraction patterns of an Aluminum-Magnesium alloy with a *10-fold symmetry*. (Coincidence?) For normal crystals, we can assign three integer values, known as Miller indices, to label the observable reflections. These correspond to the crystal's three-dimensional structure. In order to assign integer indices to quasicrystals, however, at least 5 linearly independent vectors are necessary. Only in this higher-dimensional space we can describe a quasiperiodic structure as a periodic one. The actual quasiperiodic structure in the three-dimensional physical space is then obtained as projections into that space.

As we shall see in the coming chapters, the notion of a radically enfolded, multi-dimensional space-time continuum is consistent with numerous other developments, which, in turn, support our evolving thesis of a spiritual universe.

So, do we live in a universe of marvelously intricate structure, or a vast sea of chaos? The answer, as we'll see in the next chapter, may be: 'both'.

Chaos Theory

Chaos Theory, as the name implies, deals with the chaotic behaviors observed in nature. Like the other theories presented, it is developed in a highly rigorous mathematical manner. (The branch of mathematics used in Chaos Theory is known as 'fractals'.) Basically, Chaos Theory proves that simple systems give rise to complex behaviors, which, in turn, exhibit surprisingly simple patterns. More important, the laws of complexity hold universally. Consequently, the chaotic systems responsible for the formation of a snowflake, global weather patterns, the great spot on Jupiter, electronic circuits, the spread of measles, and the formation of galaxies, exhibit uncanny similarities, which suggest a universal underlying principle.

For our current purposes, the most interesting outcome of Chaos Theory is the concept of sensitivity to initial conditions. Basically, this says, in effect, that a butterfly stirring the air in Beijing today may ultimately cause storms in Chicago next month. This differs diametrically from traditional physics, which generally assumed that such initial effects ultimately cancel themselves out.

This can be seen to have direct implications with respect to what we've already learned about quantum reality. A position espoused by most mainstream scientists is that although it is largely indisputable that observation influences events at the quantum level, the mind does not generate sufficient 'wattage' to affect the macroscopic world. Sensitivity to initial conditions, as demonstrated by Chaos Theory, provides a clue to how the weak energy of the mind might, if properly directed, influence large-scale reality (or, to be more precise, that aspect of reality that manifests itself in our 4-dimensional space-time continuum).

At the risk of getting ahead of myself, I'll observe parenthetically that this model dovetails remarkably well with reports of shamans and other occult practitioners who claim that the results of any particular practice are never instantaneous. Is it possible that by applying the power of human perception to identify precisely the right cusp at which to nudge nature, a certain outcome could be promulgated? Then again, if we wish to integrate this concept with that of the many-worlds theory, we'd acknowledge that the desired result is one among a multitude of instantiated outcomes, and the shamanistic practice serves only to direct the practitioner in the desired direction. These thoughts, however, are more suited for a later chapter. For now, we'll conclude this section with a quick overview of the Big Bang Theory.

The Big Bang

Since the beginning of human civilization, people have always questioned the origins of their existence and the creation of the universe. "Where did it all come from?" "How was it created?" These are the questions that mystified ancient societies, and they still puzzle scientists today. Cosmology, the study of the large-scale structure and evolution of the universe, is the branch of science that has evolved from these questions. Within this field of study, the Big Bang model has become the most prevalent theory. The majority of evidence from a variety of different investigations makes it extremely likely that something like the Big Bang actually occurred.[1]

The Big Bang was initially suggested because it explains why distant galaxies are traveling away from us at high velocities. The theory also predicts the existence of cosmic background radiation (the glow left over from the explosion itself). The Big Bang Theory received its strongest confirmation when this radiation was actually measured in 1964 by Arno Penzias and Robert Wilson, who later won the Nobel Prize for their discovery.

According to the theory, at the moment of its inception, the entirety of our universe was compressed into a volume the size of an atomic nucleus. As described by the prevailing cosmological models, an ineffable explosion, trillions of degrees in temperature,

[1]In 1999, the age of the cosmos was determined by NASA's Hubble Space Telescope Project Team to be approximately 12 billion years old (plus or minus 10%).

created not only the fundamental subatomic particles, and thus matter and energy, but space and *time* itself.

The universe began its existence as a dense, hot globule of gas expanding rapidly outward. At that time, the universe contained nothing but hydrogen and a small amount of helium. There were no stars or planets. The first stars probably began to condense out of the primordial hydrogen when the universe was about 100 million years old, and continued to form as the universe aged. Our own Sun arose in this way some 4.6 billion years ago. Many stars came into being before the Sun was formed, and many others were formed after the Sun appeared. This process continues, and through telescopes we can now see stars forming out of compressed pockets of hydrogen in outer space.

With this picture of an expanding universe emerging from the Big Bang, the question that comes to mind is whether this expansion will continue forever, slow to an eventual halt, or reverse itself in a process of universal collapse. The prevailing theories hold that, basically, if the universe is dense enough, the expansion will eventually reverse and contract back into an infinitely dense singularity. If the density is lower, then the expansion will continue forever.

The density of the universe also determines its geometry. If the density of the universe exceeds a certain critical value, then the geometry of space is closed and positively curved like the surface of a sphere. If the density of the universe falls below that critical value, then the geometry of space is open, and negatively curved like the surface of a saddle. If the density exactly equals this critical point, then the geometry of the universe is flat like a sheet of paper. Thus, there is a direct correlation between the shape of the universe and its ultimate fate.

These possibilities are represented by three mathematical models: The first, known as the *Open Friedmann-Lemaître Theory*, predicts a hyperbolically curved space-time continuum that is destined to expand forever. The second possibility, called the *Closed Friedmann-Lemaître Theory*, predicts a spherically curved continuum, destined ultimately to collapse back in on itself. The third possibility, known as the *Einstein-deSitter Theory*, proposes a flat-space universe that is also fated to expand.

A question that is often raised is this: if the Big Bang created the universe as we know it, what, if anything, existed before that monumental event? Obviously, if time itself came into being at the moment of the Big Bang, the question itself is essentially meaningless (a koan,[2] if you will).

One popular speculation among scientists is that the present expansion may be one cycle of many that this closed universe has undergone. (This view is shared by the Hindus, who believe the universe periodically expands and contracts, and to this galactic periodicity they give the name *kalpa*.) This, however, leads to a conceptual inconsistency. Recall, for a moment, our discussion of Special Relativity and its implications regarding time. The difficulty we encounter is that speaking of *oscillation* implicitly infers a change over time. The question we must ask is: by what chronometer might we time this universal oscillation? Therefore, if we accept that time comes into being at the moment of inception (the Big Bang), and ends at the moment of collapse (and at the center of black holes), we are faced with a paradoxical difficulty.

Another interesting aspect of the Big Bang theory has arisen in the field of computational physics. Using advanced algorithms running on Cray™ supercomputers to model the early conditions in the universe, results suggest that 90% of the original mass of the universe remains unaccounted for! Is this percentage related to the ten-dimensions required by String Theory? Alternatively, it does not seem unreasonable to speculate that this 'missing matter' may constitute the substance of those Everett-worlds of which we must remain forever unaware.[3]

Let's conclude by wrapping up our brief exploration of scientific certainties and retracing our logical steps to this new view of the world around us.

Since painful experience has proven the existence of atom bombs, nuclear radiation, and the like, we can be reasonably certain that Special Relativity is essentially correct. It, in turn, proves that the space-time continuum must be understood as a vast expanse

[2]A Koan is a paradoxical utterance used in Zen as a center of concentration in meditation. The paradoxical nature of Koans is essential to their function: an attempt to break down conceptual thought. Koans are constructed so that they do not succumb to conceptual analysis and thereby require a more direct response from the meditator.

[3]The other prevalent theory as to the fate of this 'missing' matter is that it exists in the form of so-called 'dark matter', which cannot be seen. At this point there is no direct evidence that dark matter exists, and it remains just a convenient way to explain certain anomalies in galactic motion.

in which past, present, and future are somehow embedded in the fabric of reality.

Innovations like photoelectric cells and electron microscopes prove, respectively, the particle nature of light, and the wave nature of matter. These, and many other components of everyday life, have established beyond any reasonable doubt that Quantum Theory (often referred to as the most successful theory in history) does indeed model real-world phenomena with unimpeachable regularity.

Based on Bell's proof, which has been validated mathematically and verified experimentally, we know that if Quantum Theory is true, then nonlocality must be an intrinsic aspect of our local space-time continuum. This infers that there are connections that represent routes shorter than a straight line between two points. General Relativity's proof of space-time curvature suggests one possibility of how this can be: If the universe itself is radically curved, certain interactions may somehow follow a straight line through curved space. (Remember, both relativity theories were based on the implicit assumption of a 4-dimensional space-time continuum. It may well be that these are merely a subset of a higher dimensionality, just as the 'flatland' we saw earlier exists as a limited domain within three-dimensional space.)

Although the temporal implications of Special Relativity seem, at first, to imply a disturbing picture of total predestination, Everett's many-world interpretation of Quantum Theory provides an avenue through which the concept of free will is restored. In this model, every possible eventuality exists in the space-time continuum, and through our choices, we select the ones we will actually experience. Conversely, it may be that the entire ensemble of variations constitutes the full experience of a life, and only subjectively do we appear to be experiencing only a single path through the matrix.

From mathematical models of the big bang, we discover evidence of where the substance of these alternative Everett-worlds might have originated, for some 90% of the initial matter in the universe is unaccounted for in our observations of three-dimensional space.

Finally, the evidence for the Big Bang theory further supports these other concepts. If time itself came into being at the moment of the Big Bang, and ceases to exist at a final collapse, then the entirety of the space-time universe must be envisioned as a sort of bubble of reality. But what, if anything, is *outside* that bubble? Indeed, some modern theories speculate that there may be entire universes in which time itself does not exist! As we shall see, the answer suggests a very likely candidate for spiritual, mystical, and paranormal phenomena.

PART II: HOW REALITY LOOKS

Having devoted the previous section to presenting the underlying principles of science, we now build upon that foundation. In this section we show how some of the strange concepts described in the previous section actually seem to 'look'.

Mysteries In Flatland

Let's return to our seriously constrained friends whose world consists of only two dimensions (plus time). As before, we find it useful to consider this world (Flatland) because it allows us to envision what it's like to see a subset of reality from a higher dimensionality than that within which it exists. Although it's difficult to conceive what it would be like to look in on three dimensions from a four-dimensional perspective, we can achieve a qualitative feeling for these concepts by considering a two-dimensional topology from our more familiar three-dimensional viewpoint.

The Möbius Strip

Remember what happened when we slyly transferred our hapless flat mathematician from his familiar plane environment to the surface of a sphere? At first, the change wasn't immediately apparent because to him, just like us, the surface of the sphere appeared no different than a vast plane. Nevertheless, as he started to delve more deeply into his world, various incongruities began to manifest themselves. Though 'common sense' told him that he was still living on a flat sheet of paper, eventually, he was able to deduce that his local space-time continuum was something other than flat, although he might not have been able to intuit the actual nature of the curvature. This time we'll play an even meaner trick by relocating our flat friend to the surface of a Möbius strip (see Figure 16).

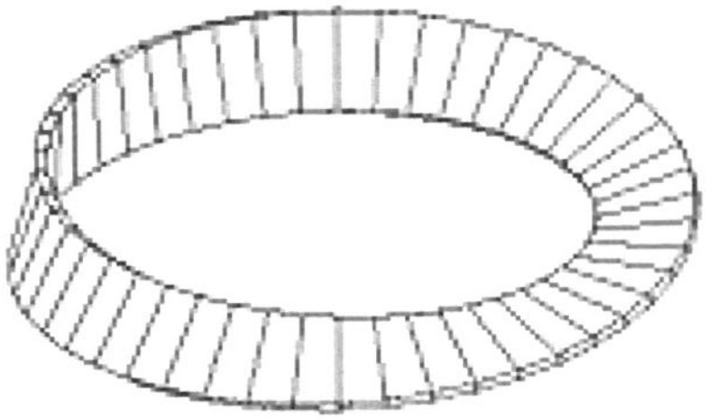

Figure 16

The Möbius strip is a two-dimensional surface that has only one side and only one edge. You can convince yourself of these properties by constructing one. Simply take a narrow strip of paper, as shown in Figure 17, and tape the ends together after giving one end a 180-degree twist; that is, connect point C to point B, and point D to point A.

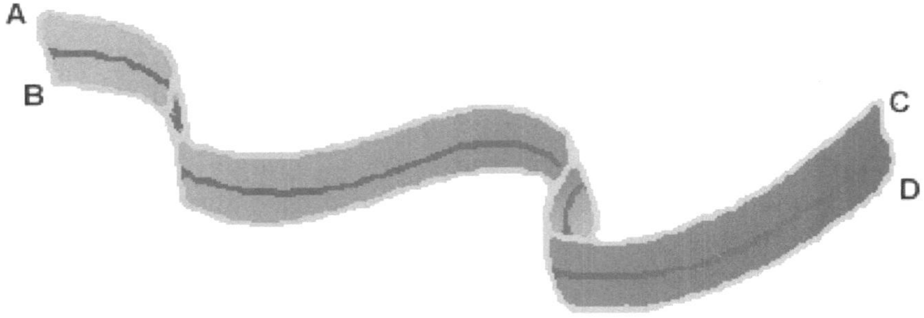

Figure 17

To convince yourself of its properties, first, start midway between the 'edges' and draw a line down its center. Continue until you return to your starting point. Although you never crossed an edge, you will find that you have drawn a line down the middle of 'both sides'. Next, hold the edge of the Möbius strip against the tip of a felt-tipped highlighter pen. Color the edge of the strip by holding the highlighter still and moving the strip across it. Although you never actually switched, every edge of the paper will be colored. As promised, you'll find that this topological structure has only one side and one edge.

Note, parenthetically, that you could have constructed the Möbius strip in one of two ways, depending on whether you twisted the end in one direction or the other. This would not have changed the fact that it has only one side and one edge, but will be significant later.

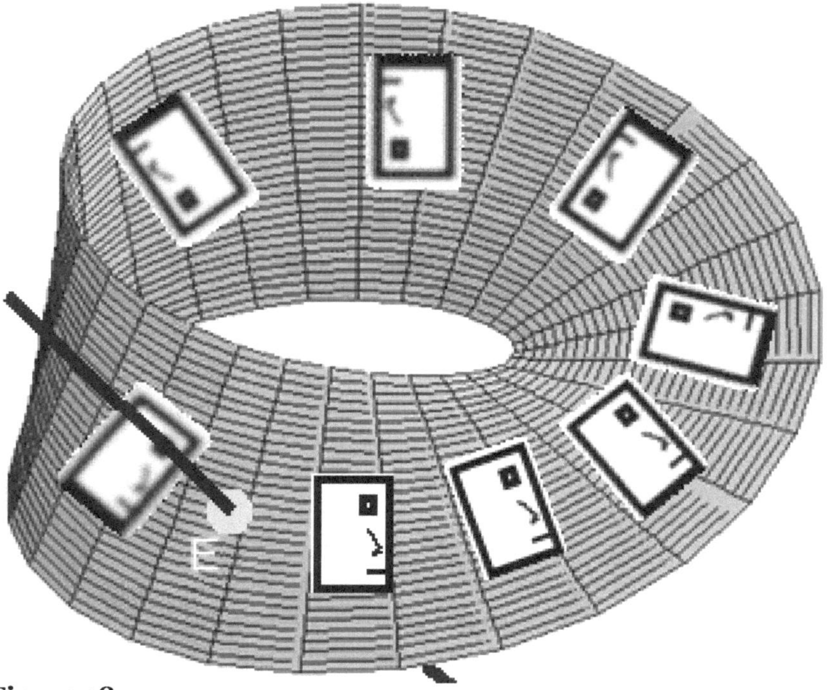

Figure 18

In Figure 18 we can see how Mister Flat might sojourn off in one direction, and return to his initial place of departure, at which point he will have explored the entirety (both sides) of his domain. Initially (if the dimensions of the strip are the same as his original sheet of reality), he will notice no difference. He will think that his world is a simple one-sided plane. Suppose, however, that we were to poke a tiny hole through the paper at point E. Obviously, the hole will also appear at another point, F, (which is not shown in the illustration). Although we can see that this is simply the 'other side' of the paper, from Mister Flat's perspective, these are two distant points on a single plane.

Now, Mister Flat and one of his colleagues, Doctor Shallow, might set up an experiment to investigate the properties of these 'two' holes. Flat and Shallow both synchronize their clocks and position themselves at these two distant areas of interest. Then Mister flat sends a message through his hole, which, of course, Doctor Shallow is easily able to receive from his. When they later compare notes, they find to their amazement that the message transfer was instantaneous! Since they have no way of knowing that the two holes are actually connected, it appears to them that nonlocality has been demonstrated, and their signal has propagated faster than the speed of light. Similarly, if they were to build a spaceship capable of squeezing through this hole, it would seem to have traveled faster than light, even though, in reality, it had taken a shortcut. This suggests a likely solution to how UFOs, if they actually originate at seemingly distant points in our universe, might be capable of traversing those distances in order to visit our planet. Similarly, it suggests a mechanism through which phenomena like telepathy, or clairvoyance might operate.

The Klein Bottle

The concept of the Möbius strip can be extended into higher dimensions, though the results are more difficult to envision. You'll recall that we observed how the Möbius strip could be created in two different orientations. Well, in 1882, a fellow named Felix Klein imagined sewing two such complementary Möbius Loops together to create a single-sided bottle with no boundary. Its inside is its outside. It contains itself. This cannot actually be done, since it requires four spatial dimensions to exist, but a three-dimensional approximation is shown in Figure 19. (Although, it appears as if the neck of the Klein bottle intersects the body, this is only an artifact of our three-dimensional representation. In reality, the Klein bottle intersects itself through the fourth dimension.)

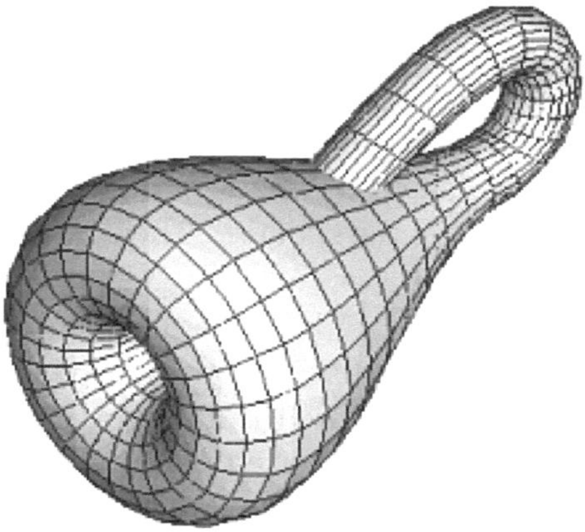

Figure 19

Although we can construct an approximation of a Klein bottle in three dimensions (Figure 20), the result is actually a projection of a four dimensional object into three-dimensional space.

Figure 20

This is analogous to illustrating three-dimensional objects in two dimensions, as shown in Figure 21.

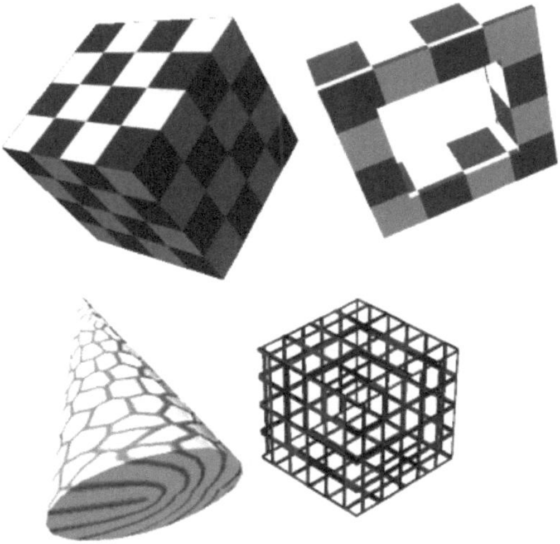

Figure 21

A true Klein bottle, however, requires 4-dimensions, because the surface has to pass through itself without a hole (see Figure 22).

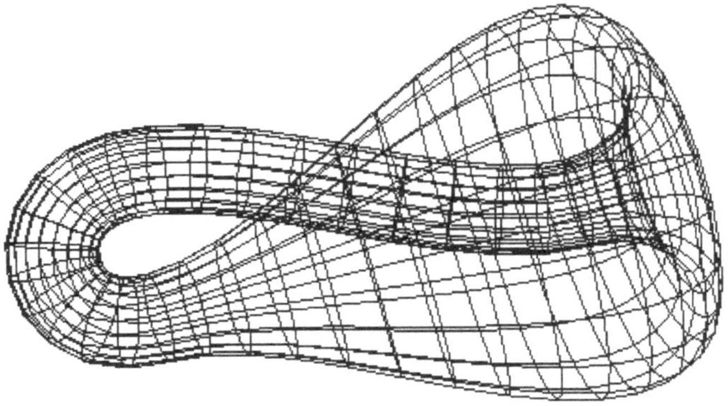

Figure 22

Compare this with a normal bottle—a wine bottle, for example. The wine bottle has two distinct surfaces: internal and external. If we were to put Mister Flat on the outside of a wine bottle, he would never be able to get drunk. Even if the cork were removed, he would not be able to pass over the lip, since this would be the same as moving over the edge of his original flat sheet of paper, which, in turn, entails moving through three dimensions. Thus, the uncorked bottle is topologically indistinguishable from a flat disk, which also has two distinct sides, separated by an edge.

The Klein bottle, on the other hand, has only one side, and hence, no edge; it is boundary-free. Mister Flat can walk across the entire surface without ever crossing an edge. Thus, the Klein bottle—both theoretically and practically—is one-sided, like the Möbius strip. It has no inside, outside, or edge.

One interesting mathematical feature of the four-dimensional Klein bottle's topology is that after each traversal of a complete circuit, our traveler would find that he has become his own mirror image. (The mathematics are too onerous to repeat here, so you'll just have to take my word for it.)

Some cosmologists, such as DeSitter, Friedmann, and Lemaître, have proposed the possibility that our Universe itself may represent a topology similar to a Klein bottle! One clue that would support such a theory would be if the same objects turned out to be visible in different directions. If, in fact, the universe is multi-connected on a scale of more than about one hundred million light-years, the only objects that could be easily seen at such distances are quasi-stellar radio sources, or quasars, for short. (Some scientists have speculated that quasars may be the 'other side' of black holes, analogous to the hole poked through our Möbius strip. In this sense, they envision the black hole as an 'input' and the quasar as an 'emitter'.)

Thus far, several candidate quasar configurations have been identified, which, after some arbitrary rotation and reflection, appear to match other configurations elsewhere in the observed universe. If these correspondences are indeed due to topology, this would imply that the universe, like the Klein bottle, is what physicists term 'non-orientable'.

Another way of considering the implications of this possibility is to rethink the Special Relativity paradox of two twins. (The two-twins paradox describes a situation in which one of two twins stays on earth, while the second goes off in a space ship, traveling close to the speed of light, then turns around and returns at a similar velocity. When the second twin gets home, he will have aged only a few years, whereas the twin who stayed on earth would have grown old and gray. The purpose of this paradox is to show the effects of the slower clock that results from near-light velocities.)

In the variation of this story that assumes a Klein-bottle universe, one twin stays on the Earth while the other goes off in a rocket, again at a speed very close to that of light. In this instance, however, the second twin arrives back on Earth without having to change direction at all. This time, moreover, the twin who remained on earth (if still alive) would be shocked to find that the returning twin's heart was now on the right, rather than the left side, and in fact, all distinguishing marks would be reversed. (From the perspective of the traveler, however, it would appear as if the Earth, and everything on it, had been mirrored with respect to the planet he'd left!)

Difficult as this model of the universe may be to envision (or believe), it would help explain a number of observable phenomena which, when examined closely, are no less outlandish. For example, the notion of black holes and quasars being opposite sides of the same phenomena no longer requires hypothesizing some nebulous 'other universe', traversed by scifi 'wormholes'. It can be seen that they actually represent a hole through the twisted topology of the same space. It also, in a sense, helps clear up the difficulty of envisioning a closed, yet infinite universe, which is harder to imagine otherwise. Similarly, phenomena like that pesky nonlocality become a bit easier to accept, as does the possibility of interstellar space flight (rather than 'going around', you instead 'go through'). And, of course, we've already seen that String Theory requires the assumption of physical dimensions in excess of our familiar three.

In many respects, this idea combines nicely with other components of our model to draw a picture of a universe that permits numerous spiritual, mystical, and paranormal phenomena in a comfortably intrinsic manner. For example, if we make the assumption (which, admittedly, is not justified by any current scientific evidence) that the realm represented by the mathematics of the complex plane constitutes:

1. The area 'outside' of our Klein-bottle Universe, and
2. The realm within which spiritual and psychic forces operate,

we may then hypothesize answers to virtually all mystical or preternatural occurrences that have long perplexed the scientific mind. Prophesy and divination, for instance, may simply involve a facility to see through the fabric of material space-time to regions which are comparatively nearby, when viewed from 'outside' the material veil. So-called hauntings might well turn out to be not, as commonly imagined, the intrusions of spirits into the here-and-now, but rather, our own voyeuristic observation, through a channel opened by the powerful spiritual upheaval, of some distant event. More importantly, such a model of the universe provides a rational structure within which to understand the 'location' of the soul. We'll explore these possibilities more fully in the next section.

The Experience of Time

As stressed repeatedly in the preceding chapters, it is crucial that we come to accept that the thing we intuit as time, and the reality of time as one component of a space-time continuum, are two very different things. Our awareness of time might be likened to the sweeping beam of a lighthouse in a pitch-dark night. We seem to see only a small circle of sea, shore, land rushing by us. The rest of the vista, however, still exists, even though we cannot see it.

Thus, although it seems as if we have control over our movement in the three spatial dimensions while time flows inexorably by, this is just an illusion. If we could behold the vista of time the way we survey the expanse of a landscape, we'd perceive that events, in reality, do not unfold before us with the passage of time. Rather, they would present themselves all at once across the field of our vision. It is merely an artifact of our peculiar relationship to this continuum that causes us to witness one narrow strip of visibility as it sweeps across the temporal dimension of the space-time matrix. Each of us, through the movement of this view-port, seems to discover new slices of existence, even though the entirety is already established prior to our awareness of it.

The philosophical difficulty with this picture is that it smacks of determinism. Without some modification, we are faced with a model in which we are helplessly swept along by a raging torrent of time, like rafters without paddles (free will). If the events of our lives are somehow embedded in the four-dimensional space-time continuum, if they simply *are*, then how can there ever be such a thing as individual volition? The answer lies in merging this tenet of Relativity with Everett's Many-World interpretation of Quantum Theory.

Of course, both theologians and scientists often have difficulty aligning intuition with the realities of time. John Polkinghorne (who is both a physicist and a Catholic priest), provides an example that typifies our tenuous grip on temporal issues when he says of God that: "Even He does not know the unformed future . . . for that future is not yet there to be known. The adoption of this view requires us to take God's temporality very seriously."[1]

As this quote demonstrates, it's all too easy to become hopelessly ensnarled by a vision that has God as helpless before the flow of time as we mortals seem to be. Nonetheless, we must understand the Universal Mind, if it exists, to be external to time itself. The physical 'thing' that came into being at the moment of creation included time itself, and hence, concepts of a God constrained by this component of the physical universe are childishly simplistic. Similarly, if time itself comes into existence at the moment of the Big Bang, then there is manifestly no such thing as what God was doing 'before' the Big Bang. Just as *He is that He is*, so too can we understand that *it is as He would have it be*, beginning to end, in a timeless act of perfect creation/will. We'll talk more about the Universal Mind and the nature of creation shortly.

Considering what we already know of the quantum nature of our local reality, it isn't too great a stretch to envision each instant as an independent, self-supporting 'object' of existence. The only connection of each moment to those preceding and those following is actually an artifact of our conscious motion from one event to the next. (This is a view propounded in Alfred North Whitehead's Process Philosophy.)[2]

Much like the frames of a motion picture, each quantum instant differs from the next ones, and from those that came before, by a small but discrete degree; yet, when shuffled by the thumb of consciousness, the illusion of continuity appears. But whereas each frame of a film has at most one other to either side of it, this 'film' is

[1]John Polkinghorne, *The Faith of a Physicist*, New Jersey, Princeton Univeristy Press, p. 81.

[2]Alfred North Whitehead, *Process and Reality*, 1978, New York, The Free Press.

like a bundle of yarn—squeezed in the center between thumb and finger, squeezed so tightly that the middle is no more than a single thread. Thus, each moment may be reached from a multitude of previous possibilities, and may be followed by a myriad of others.

From this perspective we can see how determinism is avoided, even though we maintain the model of a static, pre-determined universe. Conscious volition—free will—is the mechanism by which we choose our path through the almost endless selection of alternatives. (Or, do we take all paths 'at once', like the photon that traverses two slits in an experimental apparatus simultaneously?)

Time in the Theological Perspective

Thus far, we have based our picture of time as an independent dimension primarily on scientific arguments. It is, however, important to realize that theological reasoning leads to a remarkably similar conclusion. For example, in the fourth century, Augustine examined the paradoxical nature of time in his *Confessions*. Augustine asked of God:

"How could they pass by, those countless ages, which you had not [yet] made, although you are the author and creator of all ages? . . . Or how did they pass by, if they never were? . . . You made that very time, and no times could pass by before you made those times. But if there was not time before heaven and earth, why do they ask what you did then? There was no 'then,' where there was no time."[3]

Clearly, Augustine had a far more astute vision of God's relation to time than does Polkinghorne. Furthermore, Augustine appears to contemplate the notion of a quantized time, observing that when time can no longer be divided into increasingly minute parts of a moment, that unit alone can be referred to as the 'present'. Later, he wonders at how prophesies of the future could be explained. He examines the concepts of past, present, and future, speculating that when the present emerges from the future, it may well proceed from some "hidden source," and as it recedes into the past, it similarly goes to some other place. From this, he concludes that time is "nothing more than distention."

[3]Saint Augustine, *The Confessions of St. Augustine,* Translated by John K. Ryan, New York: Bantam Doubleday Dell Publishing Group, Inc., 1960, pp. 286-287.

Evidentially, Augustine intuitively understood that time—past, present, and future—exists in a space-like manner. Like us, he struggled with the overwhelming implications and the difficulty of envisioning such a thing.

Since it's obviously troublesome for us to envision so complex a picture, let's simplify our model a bit so we can get a better feel for it.

We'll use a game of cosmic scrabble to observe how Everett's many-world interpretation of Quantum Theory might play out in our effort to understand time:

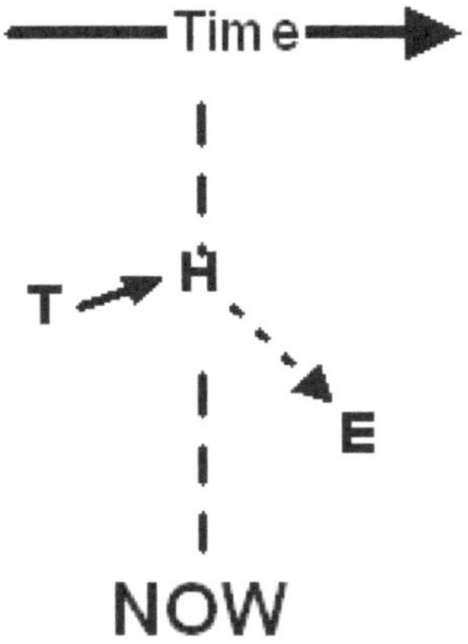

Figure 23

In Figure 23, we are in the middle of writing the word 'THE'. Let's suppose that the time it takes us to write a single letter represents the smallest granularity into which time can be divided; that is, a quanta of time. In the previous moment, we wrote the letter 'T'. We've just finished writing the letter 'H'. Since our intention is to write the word 'THE', in the next instant we will write the letter 'E'. So far, so good?

Of course, we could change our mind, and write the word 'THAT' instead, as shown in Figure 24.

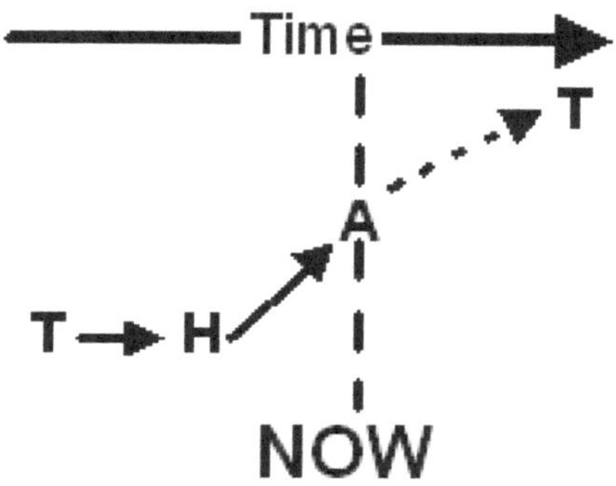

Figure 24

We just wrote the letter 'A', and an instant from now we'll be writing the letter 'T'. Life is sort of like this, right? We get out of bed, plan to have a cup of coffee, then shave, then brush our teeth; but instead we change our mind, have a second cup of coffee, then brush our teeth, and shave last of all. Nothing earth shattering. Just a little change of plan.

Now let's put a 'Many Worlds' spin on this:

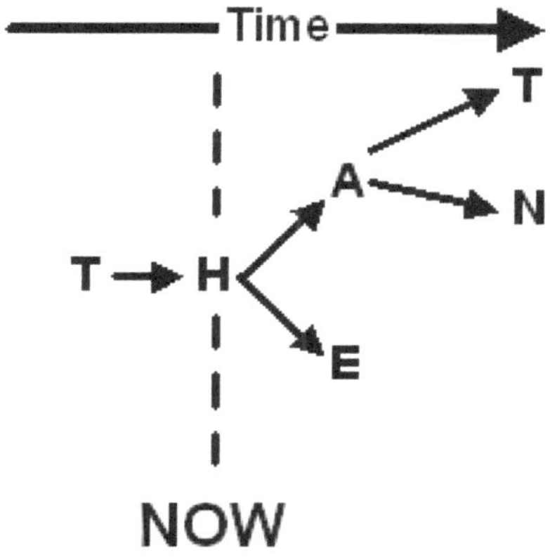

Figure 25

From our omniscient perspective outside of time, we can see (Figure 25) that there are several options open to us--we can write 'THE' or 'THAT' or even 'THAN'. Moreover, all of these eventualities are equally real. Of course, once we make a choice (e.g. Figure 23 or Figure 24), the other alternatives are no longer available. Are the other options still there? Sure, just not for us. Though it might seem as if this represents an awful waste of realities, since only one is actually 'used', we've already considered some possible resolutions of this dilemma. Perhaps we actually *do* spell all the possible words, but are only aware of a single one from within our experiential apparatus. Perhaps, alternatively, different discrete intelligences experience the same set of possibilities in different ways, as driven by their individual predilections.

Of course, words and letters don't exist in isolation, so let's see what the bigger picture might look like:

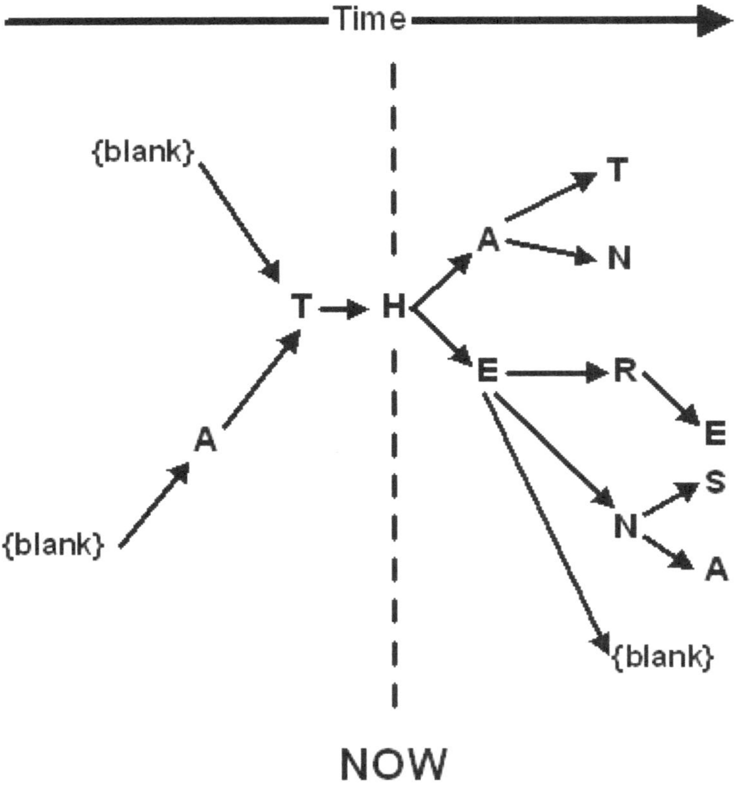

Figure 26

Now things get more interesting. In Figure 26 we see that we might have arrived at the current moment (this particular 'NOW') by a variety of routes. Depending on how we got here, not every choice will result in a sensible word. (Remember, like the game of Scrabble, we are only permitted to spell valid words.)

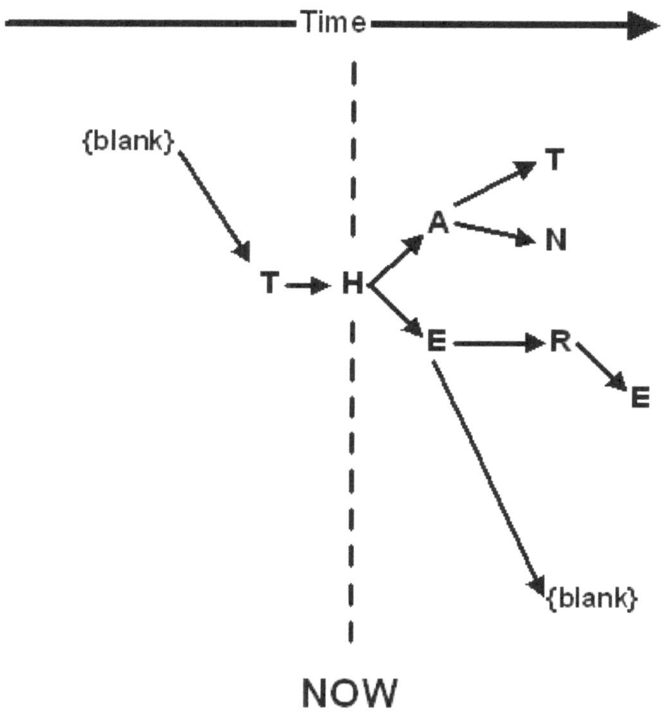

Figure 27

If we have written a blank space before the 'T', then our choices include 'THE', 'THAT', 'THAN', and 'THERE' (Figure 27).

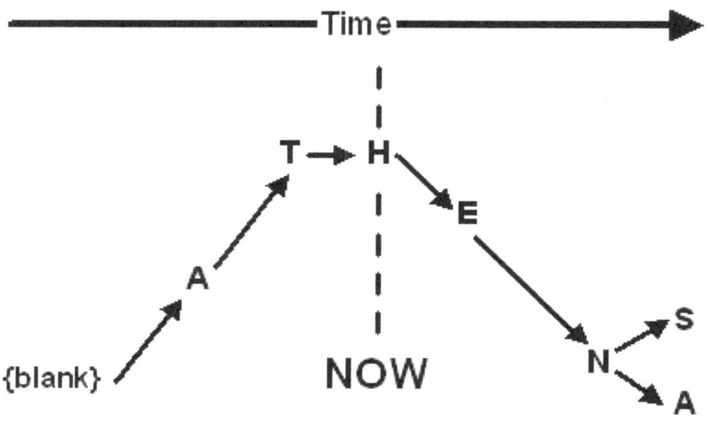

Figure 28

On the other hand, if we've just finished writing {blank}, 'A', 'T', as shown in Figure 28, then our alternatives are different: We can only form the words 'ATHENS', or 'ATHENA'.

But wait! Remember how the delayed-choice double-slit experiment proved that, under certain circumstances at least, decisions made in the 'now' can alter the course of past events? Might we, through an appropriately directed act of will, even change the foundations of the past? This is a speculation developed more fully in my short story 'Final Moments'.[4]

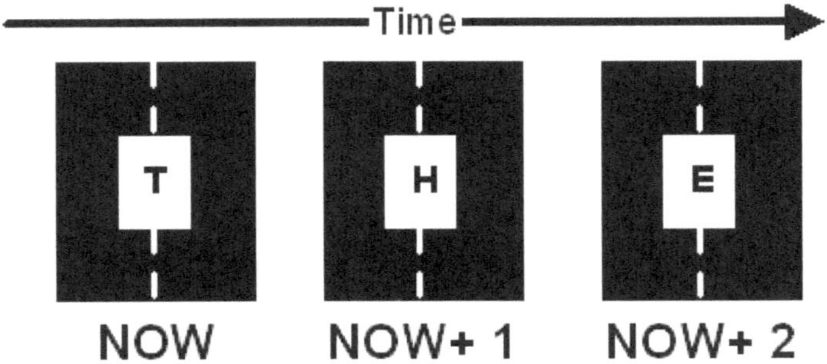

Figure 29

Of course, what we *experience* is the illusion of events developing over time. We see only a narrow window, which we think of as 'the now'. Imagine a cardboard mask with a hole cut in it, as shown in Figure 29, being slowly swept from left to right. (For the sake of simplicity we're using the case in which the word we're spelling is "THE.") This brings us back to our notion of "new slices of space-time" being revealed to us as successive aspects of the material world.

As we noted earlier, it seems kind of wasteful to have all these 'extra worlds' lying around, just in case we happen to need them, doesn't it? And what about the denizens of those alternate world lines? Are they simple automatons, devoid of life, thought, or spirit? Mere motions, like waves in space-time, with only one thread granted the dignity of awareness? From both the mathematical development of Many-Worlds, and philosophical arguments, this does not seem reasonable.

[4]J.P. Befumo, "Final Moments," in *The Space Between Time,* Philadelphia, Xlibris, 2000, p. 13

From the view outside of time, we can see how the many instantiations of a life must all have equal reality, even if they vary in probability. The 'you' who had a second cup of coffee after brushing your teeth is just as self-aware as the one who waited to brush until both cups were consumed. (They simply can't normally be aware of each other.) The question then, is whether these are distinct spirits, or collectively, the experience of a single entity. We will explore this issue in depth in the next section.

Of course, we still haven't established that there is anything more to life than this material level of self-reflection that is inherent in the Einsteinian time model; that is, each embedded moment of each variation of our lives contains its own set of thoughts, ruminations, and decisions. However, doesn't even *this* represent a very tangible kind of immortality? I would argue, however, that the prevalence and persuasiveness of some cases of past life regression, in which the individual seems to know verifiable facts about the life of another that could only have been known by the deceased, suggest that there is *some* aspect of the personality that transcends individual incarnations. Hence, we are led to one of three conclusions:

1. Spirit traverses the choices of a life, leaving the unused paths devoid of spirit, beyond that which is associated with the mechanical introspection necessary for daily life,
2. Multiple spirits may traverse the same life-tree, each taking different paths, as dictated by individual predilections.
3. The spirit absorbs the life in all of its variations, taking away a spirit-lesson represented by that life.

The first alternative I find philosophically abhorrent, for it carries with it the elitist suggestion that the majority of 'people' around us may be soulless automatons.

The second possibility has some promise. According to this model, the Mother Theresa who spent her life in service to others, and the one who went on to be an international business woman, are two distinct spirits.

The final possibility is that to the spirit, experiencing a life means experiencing it *all*—the good, the bad, the inconsequential—and coming away with some kind of lesson.

(Don't forget that 'traversal' implies a change in position over *time*. From the timeless perspective, the experience of a life is instantaneous, just the way a dream that seems to last for hours may actually transpire in mere seconds, with part of the experience being the activation of the brain's time senses.)

Let's recall our Cosmic Scrabble 'thought experiment', and see if we can make some kind of plausible sense of this. Suppose the point of this exercise isn't merely to spell *any* valid word, but to answer a particular question. For example: "Who was the Greek goddess of wisdom and prudent warfare?" In that case, the only correct path through the matrix is the one shown in Figure 30, which spells: 'ATHENA' (which just happens to be my eldest daughter's name).

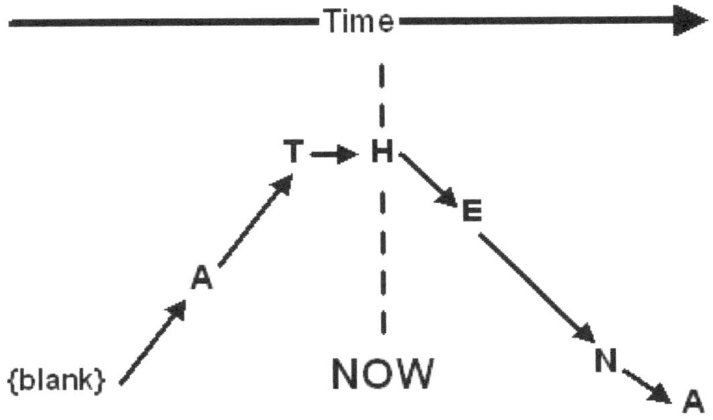

Figure 30

In the second possibility, like a multi-player strategy game, many different players may attempt the same puzzle. Once you begin, however, you must remain at a particular level until you finally solve all of the enigmas. Then you may move on to more challenging levels. Stepping back a bit, we can see how our individual lives, and all of their inherent decision points, are analogous to our simplified scrabble game. The entire matrix of lives is laid out in advance, with, perhaps, one particular path representing the ideal pattern of each existence—the lesson of that life, as it would be. The fact that only one set of options is chosen during a particular traversal no longer results in a proliferation of superfluous options.

The notion that your body may be capable of being 'used' by more than one spirit takes some getting used to. After all, we're not cognizant of possessing (or being possessed by) more than one personality, are we? Remember, however, that the overall 'self' is a combination of several components: the instincts, the intellect/mind, and the spirit/soul. Each distinct combination of physical intellect and spirit might result in a very different individual.

For the most part, these diverse versions of you will exist in different 'world lines.' Of course, there might be situations in which more than one of these entities overlap. Jung[5] observes that our psyches are composed of a plurality of loosely-knit components, not a self-contained union, and that it often takes very little to shatter this precarious illusion of a unified consciousness. Who has not had the experience, either momentary or prolonged, in which he or she simply hasn't felt quite 'normal'? Could this, in fact, represent the kind of 'parallel' event in which two world lines temporarily converge, and for a while overlap?

We'll delve deeper into the spiritual and metaphysical implications of this model in the next section. First, however, let's examine some of the geological effects that might be relevant to our emerging model.

[5]C.G. Jung, *The Basic Writings of C.G. Jung*, New York, Modern Random House, Inc., 1993, p. 56.

Geological Considerations

To some people, the notion of a connection between the planet and its inhabitants may seem no more than the ludicrous meanderings of new-age gurus, but in fact, there is sound scientific foundation for such assumptions. It's common knowledge that planets, the earth included, have magnetic fields associated with them, though fewer people may be aware of precisely how such fields are generated.

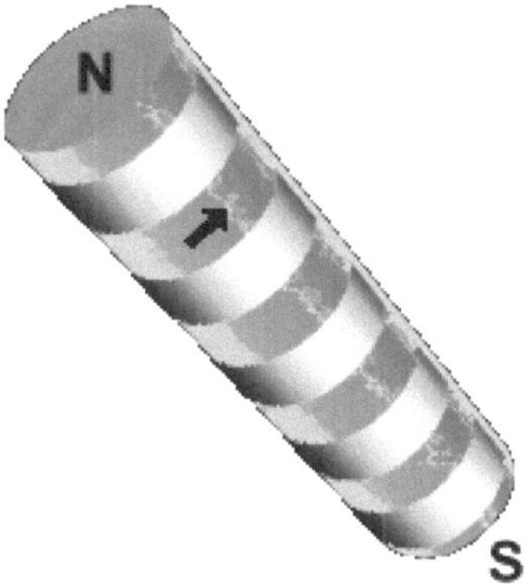

Figure 31

As shown in Figure 31, if an electric current (defined as a movement of electrons) is made to flow through a wire wrapped around a metal rod, the rod will exhibit a magnetic field. The field exhibits distinct north and south poles, as shown. (The arrow indicates the direction of the current flow.)

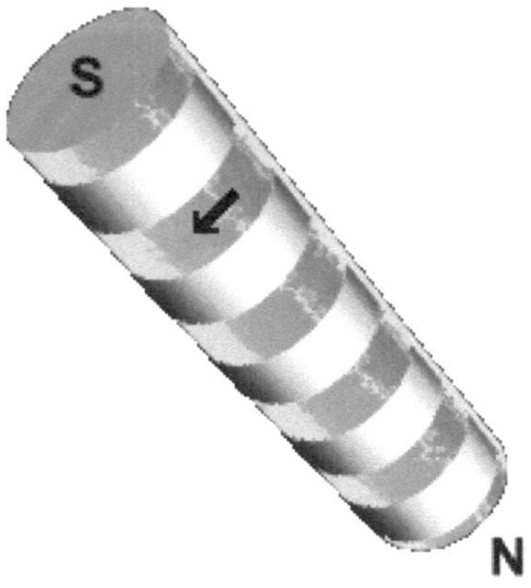

Figure 32

Similarly, if the direction of the electric current is reversed, so too will the north and south poles of the magnetic field be inverted, as shown in Figure 32.

The question that might arise is: precisely how is such a magnetic field generated in the earth? To understand the answer, it is necessary to look at the internal composition of our planet (Figure 33).

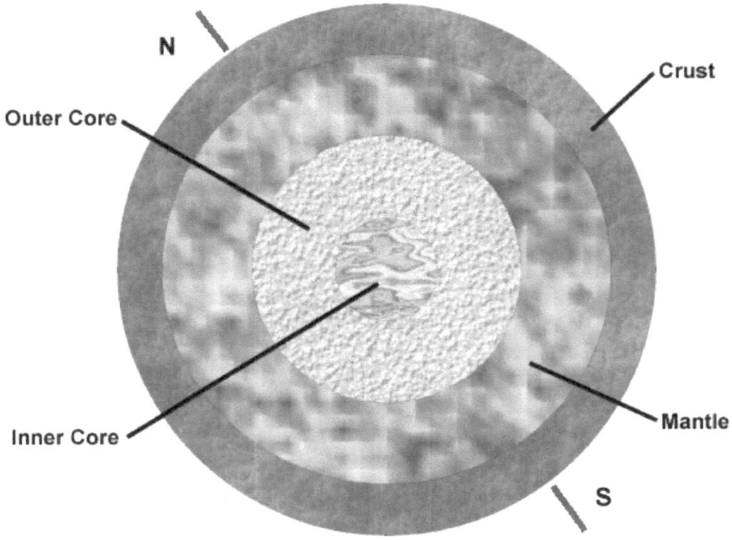

Figure 33

As illustrated, the earth is comprised of several layers (the relative thickness of the layers are not drawn to scale). The outermost layer—the one upon which we live—is known as the crust. The crust is the thinnest of the layers, ranging from 3 to 25 miles in thickness. The next layer, known as the mantle, is approximately 1,800 miles thick, and is composed primarily of molten material at high temperatures. This, of course, is what emerges in the form of lava during volcanic eruptions. Below the mantle resides the earth's core, separated into two segments—the outer core and the inner core—each existing at increasing temperatures. The primary elements in the core are molten iron and nickel. The difference in temperatures results in corresponding differences in viscosity.

Due to these differences in viscosity, the inner and outer cores tend to rotate at distinct speeds, and the variance in relative motion causes charged metallic materials to move at disparate speeds as well. This results in moving charges similar to those described in the cylindrical rod example. The effect is the induction of the earth's magnetic field.

Author William Hutton quotes William Broad as writing in *The New York Times* of July 18, 1996, that "three scientific groups [have] found strong evidence that Earth's inner core of solid iron is spinning freely within the molten outer core, making the inner core 'virtually a planet within a planet'."[1]

An understanding of these basic geophysical concepts is essential to providing credence to predictions of possible earth changes.

The Earth's Changing Fields

By examining drilled rock samples from deep beneath the surface of the earth, scientists can examine how the earth's magnetic field has changed over the course of millions of years. The reason this is possible is because metallic particles in molten material become oriented within the prevailing magnetic field of the planet, and then retain this orientation after it cools. (Picture this as a situation in which a magnetized needle is floated in a bowl of water, much like the original compasses. We might set the bowl in a freezer, in which case the needle would become fixed within the ice, reflecting its original orientation.)

[1]William Hutton, *Coming Earth Changes,* Virginia, ARE Press, 1996, p. 29.

These investigations have revealed several critical statistics.

First, it has been found that the polarity of the earth has reversed itself 171 times over the past 76 million years.[2] Second, it appears that these reversals do not happen instantaneously, but rather, follow a period in which the magnetic field gradually diminishes toward a zero-point, and then reverses itself and begins to increase toward the opposite polarity. Third, these reversals are generally accompanied by various cataclysmic upheavals. And finally, the earth's magnetic field is currently in a state of rapid decline. In fact, it is approximately 38% lower today than it was 2,000 years ago, and is currently decreasing at a rate of about 6% per 100 years. Further evidence of this effect is seen in the fact that the earth's rotation has slowed sufficiently to require the National Bureau of Standards to reset their cesium atomic clocks in order to compensate for the slowdown.

The mechanics underlying this process are not too difficult to envision.

Anyone who has ever driven a car with automatic transmission will have noticed that when accelerating from a standing stop, the vehicle gains speed smoothly and gradually. The reason for this is that between the engine and the transmission is a fluid-coupled device called a torque converter. Inside the torque converter are two sets of fins, one attached to the engine side, the other to the transmission side. Between them, the only coupling is a viscous fluid. While stopped with the engine running, the blades on the engine side turn, while those on the transmission side (which are mechanically coupled to the wheels), remain motionless. As soon as the brake is released, however, the engine's torque is transferred to the fluid, and the fluid then begins to rotate the blades on the transmission side, causing the vehicle to start moving. At first however, the blades on the transmission side do not move as quickly as those on the engine side—it takes a while for them to 'catch up'. Conversely, when you take your foot off of the accelerator, the car does not come to an abrupt halt. For some period of time, the moving mass of the vehicle causes the blades on the transmission side to continue turning at a faster rate than those on the engine side. This results in the effect known as 'engine braking', as the slower

[2]Gregg Braden, *Awakening to Zero Point,* Bellevue Washington, Radio Bookstore Press, 1997, p. 17.

moving fluid begins to retard movement of the blades connected to the transmission, and hence, the driving wheels.

It's not difficult to imagine what would happen in the earth's core if some cosmic giant were suddenly to put his hands around the earth's crust and instantaneously stop its rotation. The viscously-coupled core would continue to rotate for some period of time before it reached equilibrium with the outer surface.

As we saw when we talked about Maxwell's theory of electromagnetism, a changing electric field induces a magnetic field that tends to oppose the electric field, and a changing magnetic field induces an electric field that tends to oppose the magnetic field. The net result is a self-propagating electromagnetic wave, consisting of alternating electric and magnetic fields.

Of course, we know that photons—the particles associated with electromagnetic radiation—have zero rest mass, and as such, can propagate at the speed of light. The earth, on the other hand, has a great deal of mass, so the mutual induction effect has a far greater latency; that is, the changes happen far more gradually. Nevertheless, we can see how the relative movement of the inner core would gradually slow, come to a stop, and then slowly begin to reverse itself, thereby reversing the planet's fields. Some time later, according to some accounts, the earth's physical rotation would slow, stop, and gradually begin to turn in the opposite direction.

Biological Effects

Evidence from numerous sources suggests subtle connections between living creatures and the world in which they live. Although once relegated to the realm of folklore and superstition, science is now prepared to acknowledge that migratory cycles and patterns in various species are tied to the magnetic fields of the earth. There is evidence to suggest that magnetite, a naturally occurring magnetic form of iron oxide (Fe_3O_4), is incorporated into certain body cells in monarch butterflies, honeybees, some ants, rainbow trout, and various migratory birds. It has also been suggested, although not conclusively proven, that magnetic fields exert various effects on humans as well. It is further speculated that magnetic field sensitivity may be responsible for the ability of some animals and humans to sense the approach of earthquakes before they occur. Since the basis of our cognitive functions are known to be electrochemical in nature, it is highly likely that our electromagnetically active mental functions may be intimately tied into the surrounding fields of the environment in which we exist. In the next section we'll see how these effects may fit into our incipient philosophy.

PART III: METAPHYSICAL IMPLICATIONS

In this section we move fully into the realm of supposition. Nevertheless, it is speculation well grounded in scientific fact. Using the model of an ethereal universe derived in the preceding sections, we begin to look at traditional metaphysical issues, and explore the ways in which they are supported by the facts thus far established.

The previous chapters built an apodictic argument propounding a model of the universe that supports a metaphysical vision of reality and a spiritual approach to life. In doing so, we began with the most provable concepts--those which come closest to deserving the label 'fact', even though some may seem rather outlandish when first encountered. We must now ask ourselves whether, assuming only the most conservative interpretations of Special Relativity and Quantum Theory, we might imply any philosophy of ethics and morals, or find support for mystical tenets in such a world-view.

The Emerging Model

Let's begin by explaining precisely what is meant by a *model.*

Scientists construct models by assembling experimental evidence, associating them with mathematical symbols, and correlating these symbols in a precise and consistent manner. The model thereby constructed represents a simplification of reality that can be more readily understood intuitively because it has abstracted the vital aspects, while concealing superfluous detail. A model does not assert that something is so, but rather, it simply illustrates a particular mode of observation that entails some degree of internal consistency. The model presented in this book draws primarily upon the notions of time described by Einstein, and the quantum reality inherent in the many-worlds interpretation. If the resulting model leads to an absurdity, then we will be forced to reassess our assumptions. If, on the other hand, a convincing and consistent model emerges, then we will be able to conclude that various supported spiritual, mystical, and paranormal phenomena are not inconsistent with an objective understanding of the world around us.

Observable scientific evidence compels us to accept the reality of a universe in which clocks can run at different speeds relative to each other, and in which information can be transmitted instantaneously. We are therefore obliged to conceptualize some abstract perspective from which past, present, and future might appear as a single expanse. Indeed, we've already seen that this is a perspective required by Einstein's theory of Special Relativity. Combining the major mainstream theories of modern science (Special Relativity, Quantum Theory, General Relativity, Everett's many-world model, and Bell's Theorem), resulted in a worldview quite unlike that suggested by the illusory input of our senses. Because of the central importance of this vision, we've spent a good bit of time approaching and reiterating it from various angles, in an effort to attain a more intuitive level of comfort with this seemingly bizarre viewpoint.

From Special Relativity we've seen that matter and energy are not merely equivalent, they are actually indistinguishable aspects of one underlying phenomenon. Moreover, based on Quantum Theory, a similar observation may be applied to particles and waves. From General Relativity, we've seen persuasive evidence that everything may, indeed, be a manifestation of intrinsic wrinkles in the space-time continuum—the very fabric of reality.

In extrapolating further possibilities from our core of provable scientific tenets, the most cautious approach is to assume only:

1. The direct implications of Special Relativity: that time—past, present, and future—are in some sense an embedded series of events in the space-time fabric, and,
2. The most unambiguous assertions of Quantum Theory; that is, that nonlocality occurs under certain circumstances.

These are assertions with which even the most stodgy mainstream physicists would find no cause for dispute. Even from this minimal set of provable assumptions, we've seen that the major objections to most spiritual phenomena have largely disappeared. After all, if our actions, decisions, even our thoughts and desires are embedded with our bodies, which are components of the physical universe, it is easy to see that since time is eternal, so are we. Even if there exists no spiritual element external to what we perceive as 'the self' (and we shall argue to the contrary shortly), what remains is sufficient to represent an eternal spirit, right?[1]

Even with this minimalist interpretation, we can draw the mystical conclusion that our choices--moral or otherwise--represent an eternal 'painting' on the space-time continuum. Hence, we can argue that ethical behavior is justified, if from no other perspective

[1]Somehow, each instant carries with it the sensation of being only momentary and transitory, but this is illusory.

than by the legacy we create by our actions. From some perspective even the most hidden events are ultimately visible.

Unfortunately, as we have seen before, this simplification leads to difficulties of its own—foremost amongst them, the implication of super-determinism.[2] Thus, by considering only the most conservative implications of these theories, we are inexorably drawn to the bleak vision of a world in which free choice is merely an illusion.

Although the conservative view implies a disturbing suggestion that everything is predetermined, Everett's multi-world interpretation of Quantum Theory, combined with Bell's theorem, tells us that every possibility may actually be contained in this multi-dimensional continuum. Free will thus emerges organically as the means through which we navigate among the alternative possibilities. (But who, precisely, does the navigating? This suggests something *external* to the mundane life-line.) Indeed, we have already seen that experimental evidence supports the fact that, under certain conditions, the multiple Everett-worlds can be shown to actually exist.

The intimate connection between observer and subject has been proven in numerous instances by Quantum Theory, and is further quantified by Bell's theorem. At this point, still relying only on experimentally verified interpretations, we can easily speculate how these underlying mechanisms could support many spiritual and mystical concepts and paranormal events.

From this rather fertile starting point, we've allowed our imaginations to consider the possibility that, perhaps, the space-time curvature required by General Relativity may actually be so radical as to form a hyperdimensional topology, conceptually similar to a Klein bottle. Though not verified as extensively as the preceding theories, this, nonetheless, is more than a mere flight of fantasy. The topology suggested is mathematically sound, and there are even some preliminary experimental observations that may serve to verify its validity.

Physicist David Bohm, who has reached further than anyone into the depths of nature to draw forth a quantitative relation between consciousness and observable phenomena, has suggested that reality may resemble a hologram. (A hologram is an imaging

[2]The notion that everything that will ever occur was predestined at the instant of the Big Bang.

technique that utilizes a split laser beam to record the image of a three-dimensional object.) One salient aspect of holograms is that if any single part is projected, the result is still a reproduction of the entire image. In Bohm's model, each constituent component of the real world may similarly reflect the overall structure of the whole. This concept bears a remarkable similarity to the notion of microcosm/macrocosm, which underlies such esoteric disciplines as astrology, alchemy, numerology, and the sympathetic magic common to most shamanistic practices. Moreover, it is also supported by the discoveries of Chaos Theory, which demonstrate that the same patterns that emerge at the molecular level, are also evident in the formation of galaxies.

The concept of a radically enfolded, closed universe melds with the 'Big Bang' theory in interesting and profound ways. For example, we've already determined that, if the Big Bang is correct, then the ultimate fate of the universe depends upon the total amount of material it contains. If the mass is greater than a certain amount, the universe will ultimately collapse back in on itself; less than that, and it will continue to expand. However, if the universe is a closed system, this changes the picture considerably. Remember, at the moment of the Big Bang not only space, but time itself came into existence. It makes no sense to ask what happened 'before' that moment, nor after the hypothetical final collapse.

The notion of the universe as an eternal sequence of expansions and contractions is therefore untenable, since the very word 'sequence' implies changes over time. A more reasonable simile might be a standing wave whose endpoints are two singularities,[3] and the body of which forms our intricately folded time-hyperspace continuum. In this light, it makes no sense to ask what will be the ultimate fate of the universe. Our perspective as creatures imprisoned by an inexorably flowing time makes it difficult for us to get our minds around this notion, but the plain fact is: the universe is what it is, and does not *become* anything.

Picture this matrix as a delicate jewel; a prism within which we can perceive infinite levels of detail, forming a crystalline standing wave between two singularities, which, since they exist outside of time or space, are actually a single point.

[3]A singularity is a point of zero volume.

A Process Philosophy

In many respects, the model developed thus far is closely related to A.N. Whitehead's 'Process Philosophy' . Whitehead sees the central component of reality as the 'event'. It is the concatenation of such events that we perceive as the continuity of reality, but in fact, reality is itself granular, or quantized. Each event is bipolar (in the Whitehead model), connecting on one side (the past) to events that preceded it, and on the other side (the future) to a portfolio of possibilities to which it may lead. Human volition, in Whitehead's model, is the central 'navigator' within each event. In Process Philosophy, the events guiding elementary particles and those representing the significant decisions of human interactions differ not in kind, but only in aggregation.

All in all, this is an excellent model. It closely resembles that elaborated here, and is further supported by the discoveries of theoretical and experimental physics. The main modification I would add to Whitehead's development is one demanded by the multi-world model. Each event, I would contend, does not connect to *single* preceding event at its past, or (-) pole, but rather, to the multiple worlds that could logically have led to the event under consideration. (This is analogous to the notion of a particle having multiple quantum histories.) Similarly, on the (+) pole (future), the array of possibilities represent the equally instantiated world choices into which conscious volition may move. This interpretation is also fully consistent with the results of the delayed double-slit experiment, in which a change made to the experimental apparatus in the present apparently changes events of the past. (Obviously, the only tenable possibilities are events that might have logically led to the current situation; in other words, no change could result in a situation in which the experiment was never actually carried out.)

The important concept to note here is that Whitehead's system ultimately equates the 'quanta' of conscious volition with the elements of quantum uncertainty. In effect, this provides one more conceptual bridge between physics and psychology, matter and mind.

Yet another strength of this model is its consonance with certain schools of thought in subatomic physics, which hold that 'particles' only truly exist in terms of their interactions with others. In the same spirit, Whitehead's process components define the individual through the connections *between* events; that is, what makes you 'you' is the path selected through the myriad of possible decision points.

It is reasonable to speculate whether this vast ensemble of instants, parallel and sequential, each with its own illusion of uniqueness, collectively represent the full person, the complete you. It is you at your best, and you at your worst, and every possible permutation between. It is the Jesus we know, and the one who succumbed to temptation in the desert, and the one who used his eloquence to acquit himself before Pilot. We can imagine looking at the world-lines spawned by each and contemplating the ripples of one life in all of its possible versions, and easily perceive the lesson to be learned.

The two major objections that critics (principally, Polkinghorne) have leveled at Whitehead in particular, and Process Philosophy in general, are:

1. That the inherent 'jerkiness' produced by its central episodic nature seems incompatible with the continuous nature of consciousness, and
2. That Whitehead does not provide sufficient detail as to how low-level events are aggregated into high-level behaviors.

In particular, Polkinghorne raises the objection that, according to Whitehead, a stone, by virtue of its aggregated events, differs from an animal or a human solely in the complexity of its event aggregation.

While I can agree these are interesting questions—ones that suggest fruitful lines for future investigation—I do not believe they represent any weakness in the basic theory.

With respect to the first objection, my inclination is to restate it as one of the theory's most obvious strengths. Intuitively disturbing as it may appear, the overwhelming body of experimental evidence leads to the inescapable conclusion that we *do* exist within an inherently discontinuous or episodic reality (which Polkinghorne, purportedly a physicist, should know). The difficulty this concept presents in terms of the continuity of the 'self' is only problematic to the extent that we insist on viewing the self as an exclusively physical phenomena, and one constrained by three-dimensional space. What manifests itself as discontinuous, discreet events to our constrained perspective may well entail connections in ways too subtle for our current means of observation. Moreover, there is no reason to see this as a difficulty if we envision a spiritual basis for identity, whose existence is external to our observable space-time continuum.

Here we have the first argument for a spirit that is something more than the mere aggregate of our thoughts, deeds, and desires, as suggested earlier. These isolated events, including the material thoughts, drives, and ruminations associated with them, represent one level of immortality, however, the *driver* that actually *experiences* these events must, according to Process Theory, reside external to this ensemble of material events.

This spiritual essence is consequently seen as absorbing the ensemble of discrete episodic events, rather than sweeping through a fluid continuum (over time).

As to the second objection, it can easily be seen that this is not so much a problem with the basic theory, as a limitation in the level of detail currently available. In this sense, a similar objection could be leveled against any theory that fails to address every possibility—which is to say, all theories and models. Nevertheless, I would address this objection by observing that the model's predictions are consistent not only with our leading scientific theories, but many religious, mystical, and spiritual disciplines as well. For example, Buddhist thought has long insisted on an intimate connection between everything, whether animate or inanimate. The notion of a 'spirit' inherent in all things is common to many religions, though few will insist that the spirit of a stone is as complex as that of a human. Therefore, although we are presently unable to quantify the precise mechanisms involved, it can clearly be seen that the theory's predictions do agree with many 'observations' arrived at through independent means.

Nor should we be overly disturbed by the (currently) qualitative nature of these arguments. Consider for a moment how Darwin established and proposed an intelligible explanation of natural selection, although he remained ignorant of Mendel's experiments, and hence, had no understanding of the precise mechanisms through which genetics actually operated. Similarly, although the phenomenon of superconductivity was discovered in 1911, it took another fifty years before anyone could explain it. This, however, did not invalidate the observation. Likewise, Newton spoke with authority about gravity, though he had no clue about how it actually worked (and, in fact, we still don't). We can also compare the situation to the field of meteorology, in which, although practitioners posses a high-level understanding of weather dynamics, this knowledge remains insufficient for making precise predictions. In general, the more complex the problem domain, the further we remain from definitive, quantitative formulae. Hence, we can easily see how the second objection to process philosophy also fails to hold water.

At this juncture we have established a coherent, consistent, experimentally supportable picture of a universe far more intricate than that revealed by our senses. We have further seen that the result is fully consonant with several major systems of philosophical thought. Even without assuming the existence of a soul or spirit, we nonetheless derive a startlingly moral interpretation by envisioning the expanse of each life—every action, every decision, every success or failure—as something embedded in the fabric of space-time. The resultant 'painting' is eternal because the canvas upon which it is etched is itself timeless. In this sense, we see that life is like a Japanese watercolor, in which no corrections can be made. Every brush stroke remains precisely as it was first executed. Like a painting, the individual brushstrokes may not be separately visible, but all contribute to the finished work. We must ask ourselves: "What kind of picture do I prefer?" In the end, what you take away is an understanding of the choices and their ramifications. One may be a work of splendor, another a picture of depravity, and many simply indifferent. Nothing more can be left, nor taken away with you. Even from our constrained perspective, can we not look at the pictures created, for example, by Ted Bundy, or Richard Speck, or Adolph Hitler, and compare them to the life-art of Mother Theresa, Saint Augustine, Gandhi, or Jesus?

The important thing to remember is that there are versions in which Ted Bundy is a great humanitarian, Hitler a respected artist, mother Theresa a Bombay street whore, and Jesus just another name. Each collection of lives have a set of lessons to be learned by the spirit that chooses to partake of them.

The point at which we explicitly departed from mainstream scientific concepts was in the interpretation of the Many-Worlds model. The more mainstream view of the Universal Wavefunction has individual choices instantaneously creating one of several possible world-lines. This approach suffers from significant weaknesses, not the least of which is explaining the source of energy through which each new world is 'created'. Assuming that all possibilities are instantiated as an intrinsic aspect of a multi-dimensional universe places their creation at the same level as the creation of time itself, that is, at the origin of the universe. This interpretation also accounts, albeit qualitatively, for the mysterious 'missing matter' that appears in computer simulations of the Big Bang. Like other interpretations, this approach averts the necessity of super-determinism, with its corresponding negation of free will.

The question that arose in this model, however, was: What about all of those 'excess' worlds? It appears as if there's a proliferation of world lines that are never actually 'used', which would seem rather uneconomical.

We've already explored two possibilities. In one, freed from the constraints of linear time, distinct spirits might incarnate in different versions of the same life. In the second, it is the entirety of the life—that is, all of its variations—that represents the experience selected by 'the spirit' in material incarnation.

But what exactly *is* this spirit thing?

We have already seen that the material thoughts, reflections, choices, and apparent self-awareness of the corporeal being must be an intrinsic aspect of the time/space-embedded material reality. What you ordinarily think of as your 'self' is this physical introspection, and it in itself represents a very real kind of immortality. However, is there nothing greater?

Many mystical disciplines from around the world share a notion of communion with some greater entity. Some refer to a spirit guide, while others think in terms of communication with a personal Holy Guardian Angel. It is my intuition that the discrete spirit is greater than the aggregate of aware moments of the life, but is accessible to the cognitive apparatus of the material being though these esoteric practices.

Hence, we now see that we have one minimal scientific description, which has the entire entity as an aggregate of self-aware moments, as well as a more fanciful speculation, which requires an assumption of some component of consciousness discrete from our material selves (in all of their various possibilities). We must therefore ask ourselves whether evidence can be found to support such a hypothesis.

Supporting Evidence

Although many people may dismiss the possibility of so-called paranormal phenomena, one cannot easily write off *all* such occurrences. If even *one* such incident proves reliable, then we are forced to assume some underlying mechanism of reality through which it operates. Considering the preponderance of anecdotal evidence, there can be only three possibilities:

The first is self-delusion, mass hysteria, or the power of suggestion. Although this remains a possibility, it seems an unlikely explanation in light of the sheer number of incidents, coupled with their similarity in detail.

The second possibility is that we're dealing with something totally supernatural--mediated by capricious forces of the netherworld. In such a case, no analysis would ever be possible. Since experience suggests that we live in a world that works through predictable principles and laws, this too must be rejected on the basis of the consistent patterns that have already begun to emerge in the investigation of such phenomena.

The final interpretation--the one I find most compelling--is that these represent a manifestation of some natural mechanism, the intricacies of which we have yet to comprehend. (This interpretation does not negate the possibility that there may be a realm of spiritual entities, which may include some alien to ourselves, but only that they will most likely be bound by these extended physical laws of the universe.)

Consider, for a moment, the following:

- The governments of both the United States and the former Soviet Union have engaged in programs to investigate the possible use of so-called paranormal channels for the purpose of espionage. It is not unreasonable to assume that if advances were made in this type of research, they would not be disseminated to the general public.
- It is a widely-known fact that most or all of the major stock brokerage firms employ astrologers to provide at least some input into their strategic predictions.
- A vast number of people claim to have experienced at least one paranormal occurrence in their lives. Many of these are highly educated, emotionally stable people, for whom such an admission can have no possible advantage. Thus, it is unreasonable to dismiss all such incidents as delusional fabrications.
- Psychics have been employed by numerous police departments, and have, at least in some instances, proven successful in solving crimes.
- Dowsers are regularly employed by major oil companies and well-drillers, with a degree of success that exceeds mere chance.
- The weather-control experiments performed by Wilhelm Reich, which purportedly relied upon applications of his Orgone Energy, were widely witnessed, and their apparent success was even reported in newspapers.

Of course, many other examples can be found. These, however, should be sufficient to suggest a strong possibility that there are aspects of nature currently beyond the understanding of science, and that these may represent mechanisms underlying so-called paranormal phenomena.

The aforementioned considerations suggest that:

1. There is more to the universe than what our senses reveal, and our sciences currently understand, and,
2. It is highly likely that as new phenomena are discovered and explained, they will be consistent with the observable laws we've already found, especially considering how esoteric some of those laws already are.

Ergo, I am inclined to hypothesize that since our most widely-accepted theories of science suggest underlying mechanisms that might explain these types of phenomena, they are therefore the most reasonable places to look for such explanations. We really don't have to stretch theories like nonlocality, Bell's theorem, or Everett's Multi-World interpretation very far beyond what has actually been proven in the laboratory to support these hypotheses. Therefore, at least some of these esoteric phenomena can be assumed to be real.

As a working hypothesis, we can surmise that if

1. the fabric of space-time is radically folded, as many scientists believe,
2. action-at-a-distance is as inevitable as Quantum Theory insists, and
3. observation represents an integral part of all apparent reality,

then the influence of mind may be associated with quantum events, unconstrained by time and space as we know it.

Is such a hypothesis justified by the facts? Well, we know beyond a doubt that observation (i.e., intent) influences events at the quantum level. We know that quantum events are unconstrained by physical space (from nonlocality). We know (from the delayed double-slit experiment) that quantum events are unconstrained by time. In fact, we see from this last item that observation can, in at least one instance, exert an influence on the past. Since we began our argument by saying that if even one incident of this class of phenomena could be unequivocally confirmed, then we would be compelled to accept the reality of an entire new class of mind-linked events, the thesis is therefore proven.

One of the objections leveled against these types of arguments is that the energies involved in thought processes are of a vastly different scale than those associated with macroscopic events in the 'real world'. Clearly, the miniscule wattage generated by the brain is incapable, in itself, of (for example) bending a spoon or moving a matchstick, much less larger manifestations.

We've already encountered one possible explanation in the tenets of Chaos Theory, which states that the movement of a butterfly's wings in Beijing can ultimately affect weather conditions in Chicago.[1] Another possible answer may involve the metaphor of a

[1]However, arguments involving causality remain a bit slippery, given what we now know about time.

transistor or vacuum tube. Both of these devices operate by allowing a very weak signal to control a much larger current, producing an amplified replica of the original. Conceptually, the original signal is applied to a screen, which separates two plates, one of which provides a source of current. As the current flows through the screen, it is passed or impeded based on the polarity and magnitude of the controlling signal. The result is a more powerful version of the original signal that was applied to the screen.

We already know that we are immersed in a sea of high-energy forces. Some, like electromagnetic radiation, or the cosmic nutrino flux, are well understood. In all likelihood there are others as well. For example, Wilhelm Reich, the pioneer psychoanalyst, postulated the existence of a pervasive force known as *Orgone Energy*, which is said to represent the life force. Considering how little scientists actually know about the brain, particularly the weak fields set up by its electrochemical activity, it is certainly feasible that this physical organ might serve to control any of these otherwise undirected energies by acting in a manner similar to a vacuum-tube screen.

The brain, however, is only one part of the entire human entity. Clearly, the material organ of the brain is subject to death and dissolution as a result of the passage of time. At least part of the personality is known to be associated with the mundane operation of this physical organ. We can therefore conclude that the part of the mind associated with the brain also ceases to function with the destruction of the flesh, but we've already seen that this is illusory, for the entirety continues to exist through eternity. Moreover, the preponderance of evidence, as described earlier, suggests that there are at least some instances in which the mind operates unfettered by physical limitations. Since there is some aspect of the entity that is unconstrained by conventional limitations of time and space, it seems reasonable to surmise that it is not the component tied inexorably to the flesh. Consequently, we may reasonably conclude that there is one, and possibly several components of the sentient entity distinct from the flesh, and unconstrained by time and space. These components may correspond to the aspects variously referred to as the astral body, the soul, or the spirit. In the next section, we'll examine the ways in which numerous paranormal phenomena might be explained in light of this model.

Building upon this premise, we can draw numerous conclusions about the meaning and purpose of life that could not be justified in terms of a purely impersonal, soulless universe. Throughout the subsequent discussions of this section, I shall argue for the following conclusions:

1. That there is some part of every sentient being that exists temporarily in conjunction with the material body, but which transcends the boundaries of that corporeal self. For lack of a better word, I will refer to this as the soul.
2. That individually, these souls are instances of that aspect of God whose archetype represents the model of a life lived in its most perfect instantiation. Some attribute this model to Jesus, some to Mohammed, and others view this perfect human by other names altogether. Although I refer to this as the Christ Consciousness, this is merely a convenience, and should not be interpreted as an endorsement of any particular religion, sect, or denomination over others. It is any life in its most perfect, flawless instantiation. The soul of each human is an instance of this Christ-Consciousness in each individual. It entails a conscious awareness of the experiences the soul has come to absorb.
3. That there is a subtle connection between all such souls that is not immediately apparent within the normal limits of our corporeal sensory apparatus.
4. That the soul is the seat of some manner of individuality, although such consciousness is distinct from the mind. The soul can be associated with those drives that cannot be explained in terms of any conceivable survival benefit. Conversely, the mind is the origin of behaviors arising from an evolution-based survival instinct. Hence, the apparent personality is a combination of material intellect, seated in the mind, along with a spiritual component represented by the soul.
5. That collectively, these souls form a cohesive whole that is part, though not the totality of, a greater spirit, which we refer to as God. I will refer to this collection of souls as the Universal Spirit. It is that aspect of God who experiences the material world He has created.
6. That God does not act or create in the normal sense of the word, since action entails the promulgation of events over time, and we already have convincing physical evidence that time can not contain spirit, but rather, that spirit exists external to time, and thus,

7. That God does not create the universe from the void, but rather, the universe and God are inseparable; that is, time and the physical universe are components of a greater reality, which is indistinguishable from God. We can, however, speak of creation in the sense that it nonetheless represents an act of will and deliberate intent.

(When I refer to God, I don't intend to imply the image of a bearded Jehovah envisioned by some mainstream denominations, but rather, a universal, self-aware spirit that permeates all of creation. Be we Christian, Jew, Muslim, Wiccan, or what have you, our individual rational minds are incapable of capturing the full scope of the Universal, and hence, the concept of the Deity will apply equally to all.)

This brings us to the aspect of God that represents the underlying principle of the universe. This is the ineffable force that brings us order. Although this is the countenance of God from which all knowledge flows, it remains inherently unknowable. This is the component of God that underlies the intricate, scientific workings of the universe, and hence, represents a component distinct from that with which we engage as part of our spiritual substance; that is, it represents the cosmic mind.

One relevant observation of modern physics is that what we think of as 'empty space' is anything but. Experiments in particle physics have revealed that interactions can happen literally out of nowhere. Empty space is alive with particles suddenly appearing out of nowhere, interacting, and vanishing through mutual annihilation. So we see that what is commonly thought of as a bleak empty vacuum is far from the cold featureless void we might have imagined. Thus, a picture emerges in which seemingly empty space is animated with spontaneous creation and annihilation of particles—glittering with continual activity that could be compared to the sparkling of neurons in the brain—the thinking process of the cosmic Mind.

This sounds remarkably similar to the words of alchemist Heinrich Khunrath[2], who speaks of "fiery sparks of the world soul which is the Spirit of God." Similarly, the ancient alchemical visions of sparks scintillating in the blackness of the elemental void seem

[2]*Vom hylealischen, das ist, prae-matterialischen catholischen, oder allgemeinen natürlichen Chaos,* Magdeburg, 1597, p. 189.

echo this concept of an all-permeating cosmic mind. This suggests a universe in which the cosmic mind arises as one aspect of God—God the Universal Principle of creation—not through some mystical slight of hand, but rather, as a direct result of experimentally verified cosmological necessity.

Since we have paid so much attention to the tenets of science, it seems only reasonable to look at some of the religious and mystical disciplines, which, in essence, have been saying for many years the same things that scientists have only recently been forced, reluctantly, to accept. Since we now depart from our apodictic development of concepts fully supported by science, it seemed most reasonable to gather these more fanciful philosophical hypothesizing into their own section. Nevertheless, these items will serve to emphasize the consistency between many areas of thought, spanning the realms of science, religion, and mysticism from many lands. All are fully consistent with the scientific development here completed.

PART IV: THE LEAP OF FAITH

Faith retains a rather negative connotation in the world of science. It is often associated with fundamentalists awaiting a 'rapture', bible plaques in courtrooms, and classroom prayer. We must continually remind ourselves that all of our greatest minds achieved their distinction through faith in ideas that were not supported by the prevailing attitudes and knowledge of the day. On the other hand, everything that we are about discuss is fully compatible with the scientific foundations discussed in the preceding chapters. There is simply no scientific proof for it at this time. Nonetheless, so long as there is no evidence to the contrary, these remain viable theories.

Faith and Physics

All of the scientific theories discussed in the previous sections are, to a greater or lesser extent, currently accepted as 'proven'. We have been told to view time from some external perspective, from which the events of past, present, and future are all seen as embedded in the fabric of reality. What viewpoint is this, but the perspective of The Divinity? We are told that experiments prove that consciousness underlies the very essence of reality, and yet, many of the same scientists who tell us this deny the possibility that such consciousness might reside in something outside ourselves. After studying nature for three centuries and believing they had just about all the answers, scientists have arrived at the conclusion that mind may, in fact, underlie space-time, but they can not accept the possibility of a mind greater than that of man. How arrogant! How egocentric! They say that there are probably many universes existing totally outside of time, but are prepared to state unequivocally that there exist no intelligences that might, in some currently unquantified manner, interact with ourselves. Pretty smug, if you ask me; and what would these be but purely spiritual beings, at least from our perspective.

From a purely Socratic perspective, I would argue that, to the extent consciousness exists at all, then there exists some archetype of consciousness, and this I attribute to the mind of God; the mind of the universe of universes.

But why even postulate the existence of a God?

Jung, in exploring his theory of archetypes, observed that: "Never has [man] beheld a God, or been compelled to require the existence of such a God from the evidence of his senses. On the contrary, it needed the strongest inner compulsion, which can only be explained by the irrational force of instinct."[1] Repeatedly, he returns to the notion of an "all-pervading world soul." Hence, even if, as the atheists maintain, "man creates God in his own image," that act of conscious fabrication reflects the innate awareness of an inner spiritual component, recognized at some level of our psyche.

Jung was acutely aware of the intimate connection between so-called psychic phenomena and theoretical physics, and believed that such phenomena would ultimately be explicable in terms of a psychically relevant space-time continuum. It is my contention that one or more of the non-temporal dimensions anticipated by modern physics would be indistinguishable from what we would attribute to a purely spiritual plane.

More importantly, perhaps, Jung believed that his famous concept of archetypes actually possessed a kind of quasi-consciousness, that is, that the archetype of the soul might be a soul itself, and indeed, that which we understand as God.

The chain of logic works something like this: We begin (figuratively speaking) with "I think, therefore I am,"[2] that is, we observe that consciousness manifestly exists simply because we are capable of contemplating it. From there, we speculate that the part of the self that transcends physical time and space represents spirit, and that the archetype of that spirit is the model to which each individual aspires through the experiences of incarnation. To some, this represents the second aspect of a God-trinity,[3] to others the Christ Consciousness, but in actuality, it is an abstraction that transcends any particular denomination. Finally, the union of such spirits represents a third aspect of the Divinity that I refer to as the Universal Spirit. Again, although this formulation may bear a superficial resemblance to the orthodox concept of a "holy trinity," it merely

[1] C.G. Jung, *The Basic Writings of C.G. Jung*, New York, Modern Random House, Inc., 1993, p. 36.

[2] "*Cogito ergo sum*", René Descartes (1596-1650).

[3] This should not be interpreted as an endorsement of the Christian concept of three gods in one. There is only one God, and any attempts to delineate His various aspects are only a convenient way to help us envision something that is too great to really wrap our minds around.

represents a convenient model by which the complexity of the universal may be reasonably discussed. As a more intuitive example, suppose I were to place three fingers down on the surface of our hypothetical Flatland universe. To the residents of that two-dimensional world, it would appear as if three distinct objects had invaded their domicile. They would be unable to fathom the fact that these were but three components of a single hand. Alternatively, consider that when we look at a wave or a particle, different as they appear, they are simply different aspects of one underlying reality.

Again, although the content of this section involves notions that, at present, cannot be derived as logical conclusions of the scientific concepts discussed earlier, they are nonetheless fully consistent with them. Thus, what's required is a leap of faith, but this is a faith built upon a solid foundation of understanding, not a blind, naïve, or simplistic faith (like that of the fundamentalist evangelicals). It is a faith that the modern educated individual can embrace without apology or embarrassment.

Rene Descarte, born 1596 at La Haye in Touraine and educated by the Jesuits of La Flèche, emphasized the separation of mind from matter when he observed that "Cogito ergo sum." Advocates of Eastern philosophy take exception with this orientation, and point to it as the reason why many Westerners are aware of themselves as isolated egos, existing inside their bodies, but separated from the material world. We will, however, come to see mind as both the 'tiller' with which we navigate through the labyrinthine complexity of a space-time environment that is inherently distinct from the self, as well as the window through which the higher-dimensional self *experiences* that three-dimensional reality. That is, *mind*—the mundane thoughts, reactions, passions, and drives that define the material self—is the central aspect of the experience the soul incarnates to absorb.

This underlines the major shortcoming in the Cartesian/ Newtonian view of the world: that it separates mind from matter. A further objection would have to be its propensity to perceive the material world as some kind of complex, dynamic mechanism. As we're beginning to appreciate, the material world more closely resembles a static fixture, through which experience is navigated by the action of will.

On the other hand, discoveries discussed in the previous sections suggest that, from some perspectives, the Cartesian concept of reality as parts joined by local connections may be closer to truth than interim thinkers have imagined, providing we substitute the notion of 'mind' with that of 'spirit'. This is necessitated by the fact that Everett's multi-world vision describes the static structure of a universe that is, in many ways, intricately interwoven with the material mind (though not with spirit).

As Whitehead describes, through this complex labyrinth, individual will ultimately dictates our path. Moreover, Whitehead presents the primary motivation for choosing one possibility over another as the "lure of feeling," which is the person's "prehension of the primordial nature of God."[4] In other words, God, through the Universal Spirit, which is the collection of our individual souls, ensures that we have guidance at each crucial decision point. The lesson of life is to learn to listen to the voice of our higher aspect.

As we observed in the previous sections, Bell's theorem and Everett's model have laid a foundation for what might appear as a kind of super determinism; that is, the initial conditions can not be changed, and the universe could never be anything but what it is. Hence, it is precisely *because* spirit is distinct from the material world, and is therefore capable of exercising will in its choices among the various available world possibilities, that determinism on the individual scale is avoided.

Knowledge Versus Insight

Knowledge and insight are words typically used to differentiate between the logic of the material-centric mind, and the inspiration of the spirit. Rational knowledge consists of the abstractions gained from experience with objects encountered in "the real world." Since we've already seen how these objects are illusory, it stands to reason that knowledge based on experience with them should be considered suspect.

The experiences that led to rational knowledge were essential for physical evolution in a material environment, and hence, are not without value. Nevertheless, their limitations must be acknowledged when we begin to move toward a more sublime spiritual development. This does not suggest that analysis is superfluous.

[4]Alfred North Whitehead, *Process and Reality*, New York, The Free Press, 1978, p.189.

Discriminating thought eliminates paths that are manifestly unfruitful, constrains the imagination, and is used to scrutinize novel concepts.

Nevertheless, the aim of Zen, which is to "quiet the mind," should not be ignored. The material mind, bolstered by eons of development through physical evolution, possesses the loudest internal voice—one that tends to drown out the more subtle voice of the spirit. Hence, it represents the most formidable barrier to the true perception of reality, and it is this more acute vision that underlies the successful practice of spiritual disciplines. Moreover, even those leaps of the imagination that correspond to major scientific paradigm shifts—Relativity, Quantum Theory, the insights of Galileo, Newton, and Maxwell—all were precipitated by the *intuitive* mind.

Of course, once the leap of faith opens up a new area of thought, it almost immediately shifts into the realm of the rational mind. Relativity, for example, is now taught in undergraduate physics classes, and the techniques of applying it to relevant problems are purely mechanical. Thus, the body of knowledge that comprises the 'everyday environment' from which we draw rational knowledge is a dynamic one. Moreover, it is also one that varies from individual to individual. To the scientist, trails in a cloud chamber, images in an electron microscope, or the evidence of time dilation in a relativistically accelerated particle, quickly become part of their "everyday environment," even though at one point they represented one individual's leap of intuition.

Communicating Insights

Traditional Eastern wisdom has it that true insights cannot be communicated, but each acolyte must find his or her way to enlightenment. Nevertheless, although one may not, for example, convey the experience of gazing into the Grand Canyon, it is certainly possible to draw a map, and provide a description to ensure that the traveler doesn't mistake Pike's Peak for their actual destination. The important point is to avoid confusing these maps, descriptions, or abstract representations, and the realities with which they are associated.

Lao Tzu, in the opening line of the *Tao Te Ching*, held that the true essence of reality could never be expressed. The name he gave to this reality was the *Tao*, and he observed that the Tao that can be expressed is not the eternal Tao.

Similarly, it is my own absolute, non-intellectual experience of this reality, which can never be directly communicated, that has compelled me to write these mundane, intellect-based words, in the hope that they will help others to draw closer to that experience.

Eastern Mysticism and Western Faith

Why discuss Eastern mysticism, when this book is primarily oriented toward a Western audience? One reason is that the introspective trend of Eastern philosophy reflects an essential component in attaining the direct experience of reality. Part of the reason why Eastern mysticism sometimes appears alien to the Western mind has to do with the history of the church in the West.

From a purely objective perspective, we can state with certainty that Jesus must have said and done many things, perhaps merely mundane, which are not recorded in the four canonical Gospels. At some point, decisions were made as to which of Christ's sayings would be accepted, and which were to be concealed or suppressed. It is also widely known that the early church went through some dicey periods, when political, rather than purely spiritual interests dictated the direction that the church ultimately followed.

The Gnostics, for example, held a view of enlightenment through direct experience that is far more consistent with Eastern mysticism than with most current mainstream creeds. The reason why many such sects where condemned as heresy is not difficult to understand. Mysticism is a highly personal, internal experience, which is exactly what those who seek organizational power always attempt to dissuade. Thus, it is not so much a case of Western faith being inimical toward mystical spirituality, but rather, a question of the political body of Western churches having been motivated to maintain their monopoly on spiritual guidance. It is not difficult to see how a philosophy propounding that the only truth is that which comes from within would not sit well with organized religion.

Nevertheless, we've already seen that there exists a persuasive body of evidence that consciousness is intimately intertwined with the workings of the universe. If we further assume that this very universe is an integral part of the mind of God, then what better way (or, indeed, what other way) could there be for knowing God than by direct contact? Surely, God does not need matchmakers or go-betweens!

The Eastern Image of the Divine

In India, the Vedas are a group of four scriptures that embody the spiritual source of Hinduism. Each Veda consists of several sections, the last of which, the Upanshads, deal with the philosophical aspects of the Hindu religion and its spiritual messages. One of these, the *Brihad-aranyaka Upanshad*, describes the Divine not as a ruler, but as an intrinsic principle inherent in the universe:

"He who, dwelling in all things,
Yet is other than all things,
Whom all things know,
Whose body all things are,
Who controls all things from within—
He is your Soul, the Inner Controller,
The Immortal."

The Hindu religion often appears to worship numerous deities, but in fact, these are all considered different manifestations of one ultimate reality, known as *Brahman*. This is essentially no different than recognizing that hills and trees, lizards, and even diseases are all aspects of God. (This perspective is also similar to most modern pagan beliefs.) Just as some Western religions recognize that the human soul, though individual, is simultaneously part of God, so too does the Hindu faith recognize that Brahman is the substance of the human spirit. That part of Brahman that is instantiated in individual humans is called *Atman*, and corresponds to what some refer to as the Christ Consciousness.

Lila means "the play of God," and refers to a process whereby Brahman becomes the world, which in turn, ultimately returns to Brahman. Hinduism envisions the world as a stage upon which this divine play, lila, is performed. It is not difficult to see that this concept of the relation between Atman and Brahman is very close to many Western mystical concepts. Although the notion of lila as a play might seem to trivialize the situation, this is not so, because it is the play through which God is completely realized, fulfilled, and completed.

The word *maya* refers to the process through which Brahman transforms himself into the world, thereby instantiating the creative process. This is synonymous with physical incarnation, which in a sense can be thought of as a school in which the spirit is perfected. Under the spell of maya, the illusion of the material is presented as a 'stage of flesh', upon which the life's dramas are played out. Maya is the state of illusion within which man remains unaware of his true nature.

The Hindus maintain that ultimately, reality can not be grasped with concepts or ideas, but must be experienced directly. (This echoes Lao Tzu's assertion that the Tao that can be expressed is not the eternal Tao.) The name they attach to this ultimate reality is *sunyata*, which means 'the void'. However, in view of what we've already learned, perhaps a more useful concept might be to think of it as 'the fullness'. Only our mundane concerns, considerations, and egos obscure, eclipse, and exclude the rich content, turning it into a bleak conceptual vacuum.

The Christ-Consciousness

Love/compassion is the highest emotion, and Jesus (the version with which we are familiar, once we strip away all of the bogus silliness added by two centuries of corrupt church orthodoxy) is one example of a human who is said to have manifested perfect and complete love, while overcoming fear and all of its negative manifestations.[1] We might reasonably apply the spirit of Lao Tzu's

[1]It is important at this juncture to reemphasize the distinction between Jesus the man, and *The Christ*, which I describe as the spiritual model for a perfected human life. I use the instance of Jesus as an example, because this is the theological model with which I am most familiar. Nevertheless, the underlying concept should be understood as meaning that there exists for each of us a model for how our lives should be lived, and, as elaborated in this book, each of these can be interpreted as embodying the spirit of *The Son*. Hence, the concepts apply equally well regardless of the name by which one chooses to relate to this spirit.

observation regarding the Tao to the concept of Christ. We can speak of the life of Jesus (or Mohammed, or Buddha, or Shango), analyze their teachings, and discuss the nature of perfect being, but to appreciate the essence of this spiritual archetype, it must be experienced directly.[2] All of the other things are merely 'maps' that help bring us to a point from which we can experience perfection directly.

These days we needn't search far to find people proclaiming their "personal relation with Christ." Some, perhaps, will have actually attained the union with the universal Christ-consciousness, or nirvana, or awakening, or whatever name you care to attach to this experience. It is highly likely, however, that the vast majority merely mutter the words as some kind of incantation. Many hear the words, but few can walk the path to enlightenment, or, in the words of Matthew (22:14), "many are called, but few are chosen."

Jesus and other spiritual figures like him reached their full potential not because they were perfect to begin with, not because they were born God, but because they opened themselves fully to the Universal Spirit. Jesus, for example, encouraged his disciples to use the same source of power, and even told them that "he that beleiveth in me, the works that I do shall he do also; and greater works than these shall he do" (Luke 14:12). The intent of the Universal Spirit, therefore, is to help us unite with God as Jesus did; that is, allowing our material minds to became one with our higher selves.

The Chinese have long asserted that the person who is in harmony with the Tao, or spirit of the universe, and who directs his or her actions accordingly, will inevitably succeed in every undertaking. It's not too difficult to see that this was almost certainly the source of Jesus' power, and that this power, as he revealed, is also within our reach.

When we explore the nature of the soul we will show that this is actually the aspect of the human spiritual archetype that dwells in each of us. Collectively, these souls represent that aspect of God referred to as the Universal Spirit. Thus, as our spiritual model emerges, we are beginning to see that one aspect of God represents the unknowable creative Principle underlying the universe; another, the archetype of the perfected human spirit; and the

[2]Jesus himself observed that: "I am the Way, the Truth, and the Life. No one comes to the Father except through me." (John 14:6).

Universal Spirit, the union of all individual souls.

As stated earlier, the concept of a "Christ-consciousness" refers to a certain universal prehension of an archetype of the human spirit. As such, it is not unique to any one historical figure, but represents our inner recognition that there exist certain individuals in our history or mythology who lived their lives in a manner that approached perfection. Great wisdom, compassion, and miracles are an integral component of these figures' lives. In the following discussion I use Jesus as an illustration, not because I perceive him as being necessarily more representative of the concept, but simply because the documentation of his life offers readily accessible and widely recognized examples.

There can be little doubt, nor would current orthodox Christian denominations disagree, that Jesus was fully human. We are told that he was baptized, therefore acknowledging human birth tainted by the original sin purportedly common to all humans. He was tempted in the desert; the divine would never be subject to temptation. Even in the end, dying on the cross, he said: "I thirst" (John 19:28). The point of his humanity is significant, since resurrection after death would be a trivial feat for one who was not human, but rather, an embodiment of pure divinity.

On the other hand, a central tenet of all mainstream Christian denominations is that Jesus was not *just* a man. The primary distinction hinges on the resurrection.[3] It therefore behooves us to examine the whole notion of the resurrection, with an eye toward understanding whether such an event can be squared with our current knowledge of science, and in particular, with the model we have so painstakingly assembled thus far.

Let's begin by considering the story of Jonah and the whale. To most of us today, it would seem patently obvious that Jonah could not have survived a voyage in the belly of a whale. Moreover, it seems safe to assume that even in ancient times, the majority of the Bible's audience undoubtedly possessed some knowledge of the basic physiology of a fish's (or marine mammal's) digestive system. Consequently, it is highly probable that the Old Testament authors

[3]The raising of the dead is not an ability exclusively attributed to Jesus. The magician Apollonious was reputed to have raised the dead, as well as performing many other miracles similar to those attributed to Jesus. In fact, Apollonious stated that physical ascent into heaven was the true test of deification, the ultimate goal of magic.

intended that their readers would focus on the inner meaning behind that story, and understand it as a fable. In the case of Jesus, however, we can reasonably conclude that precisely the opposite was true.

Although stories of individuals raising the dead were not unheard of, it seems improbable that proposing such a story would have been seen by the authors of the Gospels as enhancing their overall credibility. Hence, it is unlikely to have been a conscious fabrication. Moreover, the Romans and others of that era were comfortable with the notion of ghosts and specters appearing in human form. If the writers were engaged in a strategic effort to foster a political organization, it would have made far more sense to propound a more credible assertion than that of a physical resurrection. This seems particularly reasonable because appearance in spiritual form would arguably have served equally well the purpose of demonstrating that Jesus was somehow 'special'. Thus, we are compelled to consider that, at the very least, the apostles were themselves convinced they had witnessed something so truly miraculous that they had no alternative but to proclaim it exactly as it happened.

This presents some difficulties with respect to concocting a plausible explanation in terms of modern knowledge. We can accept the notion of individuals being resuscitated, using modern medical techniques and equipment, within as much as thirty minutes after apparent death (i.e., after the heart has stopped beating). This situation, however, is radically different from restoring life three full days after death—particularly when death resulted from the kind of trauma inherent in crucifixion.

Moreover, the accounts of Thomas preclude any possibility of mistaken identity. Therefore, we are compelled to believe that the apostles did encounter some manifestation of an individual with whom they were all intimately familiar, whom they were all convinced had been put to death, and who bore every indication of being physically present and quite alive. Furthermore, with the perspective of hindsight, it is easy to see how only such an occurrence can account for the durability of the resulting cult after two full millennia.

Even if we accept on faith that this actually occurred, the related question comes to mind: where is this corporeal entity now?

Both reason and the scriptures themselves may provide some hints.

To begin with, the gospels reveal that when Mary Magdalene first encountered Jesus (John 20:14), she did not immediately recognize him, but supposed him to be the gardener. This suggests that while there was nothing supernatural about his appearance, there was nevertheless *something* different about him. However, Jesus admonishes her to "touch me not: for I am not yet ascended." (John 20:17).

These strike one as details that would neither be fabricated by the original authors, nor added by subsequent revisionists, since they do not serve to substantiate the basic claim that Jesus had risen bodily from the dead. On the contrary, they would seem to cast doubts in the first case on the identity, and in the second on the physicality of the individual Mary had encountered. These considerations argue strongly for the veracity of the accounts as recorded, and begin to point in the direction of what might actually have transpired.

Research in Kirian Photography, pioneered in the former Soviet Union, offers convincing evidence of the existence of tangible auras associated with living entities. Wilhelm Reich, the pioneer psychoanalyst and protégé of Freud, called this universal life force orgone, and believed that it could be measured and quantified. Modern Theosophical thinking associates these auras with something called the astral body. The astral body is said to correspond to man's state in the Garden of Eden, prior to his fall into total materialism. The astral body is said to contain the same internal organs and outer sensory apparatus as the physical body with which we are familiar. It possesses the powers of sight, hearing, taste, smell, and touch. It has a complete system of nerves and arteries for the circulation of astral fluid, which is to the astral body what blood is to the physical.

Could this notion of an astral body be used to explain the events reported in the gospels, and if so, how does it square with our emerging model of the universe?

To address the first question, the description of an astral body seems to fit quite well with the events recounted in the gospels. It has substance, and could thus be seen by others, as described in Luke 24:39, where Jesus invites the disciples to handle him and convince themselves that he has flesh and bones. It has organs, and we see in Luke 24:43 that the risen Jesus eats in the presence of witnesses so they can see that he is not merely an apparition devoid of substance.

Can we find room for something akin to the astral body in the recent discoveries of physics?

Well, many new-age writers forecast a coming change in which they speculate that humankind will ascend to a higher level. A popular prediction is that this will happen as a result of the enlightened somehow increasing their 'vibratory frequency'. As we have described earlier, all quantum waves are essentially oscillatory phenomena, which possess a complex or mathematically 'imaginary' component. Thus, it seems plausible that so-called astral-body phenomena might be related to a shift into this (mathematically speaking) complex state.

As mentioned earlier, Wilhelm Reich believed he had isolated this substance, corresponding to what psychics call bioplasma. (Unfortunately, all of Reich's books were ordered burned by a United States Circuit Court, and he was thrown in jail, where he later died under suspicious circumstances.)[4] Hence, although we have no direct proof for the existence of the astral body, there appears to be at least some anecdotal evidence for its reality. If this is so, then the tenets of Quantum Theory certainly offer potential mechanisms for its explanation. Or, could Jesus have somehow transcended the quantum injunction against communication of the Everett worlds, and simply manifested a version of himself that had survived his ordeal?

Thus far, we have come up with a credible explanation of how Jesus could have been fully human, and yet, might have transcended death to a sufficient extent to appear, risen, to his disciples—not as an insubstantial spirit, but as something *resembling* flesh and blood. Once again, although the orthodox try to cast Jesus as totally unique, and somehow, the Son of God in a manner unique to himself, his own assertion that others could do everything he did contradicts this interpretation. Hence, it makes more sense to look toward Jesus, and others like him, as examples of what we might achieve if we could only manage to align ourselves with the Universal.

In the next chapter we will expand upon the model in which we—that is, our perfected souls—collectively represent one component of the substance of God, which we refer to as the Universal Spirit. Just as we understand one God to have many distinct

[4]J.P. Befumo. *The Oranur Conspiracy,* Philadelphia, Xlibris, 2000.

aspects, so too does participation in the Universal Spirit represent unity without the sacrifice of individuality. Why insist that individuality be retained? Well, although ultimately we can say that we're all part of the Divine Whole, this is probably as useless an analogy as stating that the entire internet is nothing but a bunch of zeros and ones—yes, it's true, but it involves looking at things from a scale that is not particularly useful or informative. Personally, I find it hard to believe that God would give us this thing called individuality, only to take it back, forcing us, in effect, to sublimate into His substance like raindrops into the sea. On the other hand, I find it equally difficult to envision us as discreet winged angels, each walking around on clouds with our golden harps, occasionally bumping into each other. I think that somehow, the answer is somewhere between these two poles. We are each discrete, unique individuals, but it is a nature that is open for all to experience in a very intimate way. (It is worth reiterating the similarity between these multiple aspects and the wave/particle duality; each is distinct, and yet, are different perspectives of the same unity.)

From the by-now familiar hypothetical viewpoint outside time and space, we can see the entirety of history at a glance, spread with countless lives, each extending into a web of innumerable possible paths, intertwining with many other lives around it. Through each will be one path that represents the best possible set of choices for that corporeal existence. We can see how some lives may exert more or less of an effect on a greater or fewer number of lives around them, with the influence of some extending far beyond the temporal limits of their actual duration. Thus, as we regard the landscape of eternity, our vision is inexorably drawn to a few central points brighter than all others. These, of course, correspond to these great teachers, whose influences reach out through centuries to touch and inspire multitudes of others.

If you have read this far and at least nodded to the possibility of some of the concepts presented, then you are probably cognizant of the notion of enlightenment. Certain practitioners of various esoteric practices can, for brief periods, experience directly the vista of reality that is denied to conscious ratiocination. Typically, this ability comes only after the practitioner learns to quiet the ego, and opens the mind to the direct Will of the Universal. What might the life of someone who had fully mastered this ability be like? What if, at each juncture, each temptation, that person was able to step back and view the proffered choices in the light of the ultimate pathways to which each option might lead? To such an individual, the 'alternatives' would not be instantiated, because there would be no possibility of taking a wrong path (or rather, their probabilities would be infinitesimally small).

These are the lives that touch all others. As humans, therefore, they represent to their followers the archetype of what the perfect human could be, and as such, are the models of what God had in mind for man to become. Indeed, perhaps these lives represent the terminal incarnations that all spirits must traverse in their quest for perfection. Perhaps when Jesus said that "nobody comes to the father except through me," (John 14:6) he spoke literally rather than figuratively.

At the very least, we need to open ourselves to that vision in order to actually touch the eternally existent Christ-Consciousness reflected in these great examples. Only by so doing can we fully appreciate the technique through which we can live our lives as they lived theirs. This is how we may attain our final transformation; our final ascension. And how can we achieve this? By opening ourselves to the Universal Spirit; that is, the voice of our soul—the part of us that participates in the universal—at each decision point, and guiding our actions accordingly, just as Jesus and the others like him have done.

The career of Jesus exemplifies the message of life in a particularly succinct manner. By all human standards, he was an abject failure. He was cut down in his prime, sentenced to the most humiliating of deaths. He was apparently abandoned by the God he had served so faithfully ("my God, my God, why hast thou forsaken me?"). Yet, he is still seen to triumph in the end. Like Job (of whom we shall speak shortly), we come to understand that the human perspective is incapable of appreciating the true significance of material events, and that what seems to us to be evil or misfortune may actually represent our greatest good. Thus, the cross—by all normal measures a symbol of defeat and ignominy—has becomes to many an enduring icon of triumph and the reconciliation of mankind with God.

It's essential that we come to appreciate the close kinship and spiritual similarity we share with these great spirits. Despite their prominent positions in our history and mythologies, they nevertheless remain human souls, and as such, participate with us in the collective Universal Spirit. This, historically, was one of the main points of contention between the orthodox Christian denominations and the Gnostic sects. Tertullian, one of the early church fathers, condemned the Gnostics for "a spiritual and idealistic overestimate of the self which blurs the fixed limits that separate the creature from the deity."[5] Clearly, the early church did not want man to have the audacity to assume that he actually *is* created in God's image. If he did, there could be no justification for his subjugation by the political authority of the church. The fact that Jesus proclaimed that *anyone* could do the works he had done was conveniently ignored.

It's critical to understand at this point that we are not suggesting that since our souls are parts of the Universal Spirit, and our incarnations seem less then perfect, that the Universal Spirit (and, by extension, God) is imperfect. This is emphatically not the implication. Rather, our souls *are* perfect, and that perfection is manifest in the working through of our incarnations to perfect the exercise of free will. That's our job, so to speak.

It's also important to recognize that some part of our spirits remain in abeyance while we work through the lessons of our physical incarnations; that is, there is some part of the soul that does not get shoehorned into these physical shells. Indeed, it is through these external components that we participate in the divine.

By understanding that each of us is possessed of a perfect soul, which participates wholly in God as a component of the Universal Spirit, we achieve a far better appreciation for the humanity of these great religious figures. Just as we are tempted, know doubts, experience pain, so too did they, and by so doing, provide for us examples of how it can be done—how it *should* be done.

We'll return to this topic later. For now, however, let's continue our exploration of feasible conjectures by examining how concepts of spirit and creation can reasonably be extrapolated from what is scientifically proven.

[5]Kurt Rudolph, *Gnosis, The Nature and History of Gnosticism,* Edinburg, T&T Clark Limited, 1984, pp. 13-14.

Spirit and Creation

A primary conceptual incongruity inherent in traditional notions of creation is that they almost all convey an assumption of works performed over time: *first* there was God, *then* God created 'X', 'Y', and 'Z'. *Next*, He saw that it was good, and *finally*, He rested. This quaintly orthodox view would certainly have represented an intuitive explanation to early bronze-age audiences, but seems hopelessly simplistic in view of current scientific knowledge. The most obvious problem with such arguments is that they make the Deity subject to time, whereas it's more logical to assume the inverse to be true. Thus, an aspect of nature (i.e., time), which man is already capable of understanding, envisioning, and even (to some extent) manipulating, is implied to represent a limitation within which God must operate. If we're ever going to attain a greater appreciation of God, and His relation to us and the world in which we live, we must stop extrapolating our own limitations to Him. The boundaries of time are chief among these constraints.

God *is*.

When He spoke to Moses he introduced himself as "I Am that I Am." Was this just a catchy phrase, or did he really mean it? I'm inclined to assume the latter. God is God because there is nothing else God can be.

The world, according to Polkinghorne, is separate from diety.[1] What an odd assertion! Are we then to believe that God's

[1]John Polkinghorne, *The Faith of a Physicist,* New Jersey, Princeton University Press, p. 73.

creation suddenly ceased to be any part of Him? Does God then sit somewhere on a cloud, looking down on this thing, from which He is now fully divorced? Of all the fabrications of the Western mind, perhaps the most onerous is this propensity to segregate every thing from everything else, and here we see this unfortunate mindset perpetuated to its illogical extreme.

None of this should be interpreted as suggesting that the substance of the universe is *not* a result of Divine Will. Creation, however, cannot be the step-by-step cookbook procedure we're tempted to envision. Although the totality of God exceeds the substance of the physical universe, reality is necessarily an attribute of God, for otherwise it must be external to Him, which is unthinkable.

Polkinghorne seems inordinately perplexed when considering the ramifications of Process Philosophy. He sees it as demanding that we are either in the "unrelaxing grip of a Cosmic Tyrant, or the impotent indifference of a Deistic Spectator."[2] Although it would clearly be hyperbole to claim that the answer is simple, it is nonetheless not all that difficult to speculate as follows: God is unconstrained by time, unconcerned with space, immersed in a nontemporal void that we cannot reasonably comprehend. In a deliberate act of volition, He moves upon the void and creates. Time and space are created, but of course we can only 'see' back as far as the initial moments in which time came into being. We must remind ourselves, however, that God Himself is outside of time. What we perceive as the unfolding of fortuitous events in a causal order, over immense periods of time, is, to God, an instantaneous act of completion.

In His creation, God has provided us with every freedom and possibility, and thus, cannot be viewed as a "Cosmic Tyrant." Nevertheless, neither is He an "impotent observer." There is a right way for each of us to traverse the matrix—the way He wants us to live—but the other possibilities must be there if the experience is to have meaning and substance. God does not peer in at crucial moments, for to God, all moments are spread before Him, fully manifest. He makes Himself available through the Universal Spirit at every decision point, not as a coercive policeman, but as a kindly advisor. God never forces His guidance upon us, but makes it ever available,

[2]John Polkinghorne, *The Faith of a Physicist,* New Jersey, Princeton University Press, p. 80.

if only we request it.

Angels and the Soul

Angels are traditionally described as messengers from heaven. Thus, it seems reasonable to speculate that they are, in fact, the discrete souls who speak God's truths to the physical entities in which they are incarnated. Indeed, nowhere in Genesis are we told that God creates the angels. Thus, we may interpret the creation of man, "in God's image" (Genesis 1:26-27) as the differentiation of one aspect of God through the creation of the human *souls*; that is, the angels. Only later, in Genesis 2:21-22, do we see the *body* of man becoming the corporeal male and female entities with which our materially-oriented egos associate the entirety of the self.

As described earlier, these spirits are honed within the envelope of time, through the exercise of free will, by incarnating in corporeal bodies. Just as there is more to the Christ than that which we know as Jesus, it seems reasonable to assume that not all of a spirit is actually constrained within the shell of flesh. Hence, we derive the fully consistent concept of a guardian angel, or the personal deity—the 'rest of' that particular spirit that is associated with a specific group of incarnations.

From these considerations, it should be apparent that the Universal Spirit does not view us through the eyes of judgment, but rather, in terms of the ways in which it can help us connect with God. Interestingly, psychic Edgar Cayce refers to The Trinity in terms of "God the Father, God the Teacher, [and] God the Director, in the minds and hearts of men."[3] This passage seems to affirm that the thing in the human entity that directs our higher nature—that is, the soul—is one with the Universal Spirit.

The Soul and its Relation To God

We stand before a critical cusp at which religion, philosophy, psychology, and physics have converged, and the point of their

[3]Edgar Cayce, 3976-15, January 19, 1934.

confluence appears to be the spirit or soul. We can easily see how the emergence of the soul as a verifiable cosmological entity, with specific identifiable properties, may indeed shake the foundations of both physics and theology. Combining the orthodox perspective with what we have speculated about the nature of the universe, we find convincing concordances: God created man, motivated by a desire for an object of His love. Since God is all, everything is an aspect of God. Nevertheless, He wanted the soul of man to be distinct, discrete—capable of accepting and returning that love—so man is endowed with free will, and placed in a material environment within which that will can be exercised. Only thus can the result of choosing the path of righteousness be more than mere façade.

Collectively, the souls of men form the Universal Spirit, or what some refer to as the Holy Ghost.

This notion of associating the collective 'world soul' with the Universal Spirit is not original. A similar concept was proposed by Guillaume de Conches (1080-1154), a Platonist scholar. Similarly, Paracelsus says: "But if we are to describe the origin of the inner man or body, mark that all inner bodies be but one body and one single thing in all men, albeit divided in accordance with the well-disposed numbers of the body, each one different. And should they all come together, it is but one light, and one reason."[4]

If the soul of man is a part of God, one might wonder, then why are we not perfect? This is a mystery that is difficult for us to comprehend. Wise men, philosophers, prophets, and scientists have struggled with these concepts since the beginning of civilization. The difficulty, once again, comes from the way we think about time. We think about everything in terms of linear causality. Every experience we've ever encountered can be lined up in an orderly fashion, based on when it occurred. The material world exists inside of time, as we've already discussed, which is why that's the only perspective we can easily comprehend. But God is not limited by the material, and time is, in a sense, inside of God.

The spirit of man is pure, holy, and blessed, but perfection cannot be attained without trials, hardships and adversity to surmount. For this reason, God created flesh, and the world of flesh, so that the human spirit could be perfected through experience

[4]*Philosophia sagex*, ed, Huser, Vol.X, p. 341 (Quoted in C.G. Jung, *The Basic Writings of C.G. Jung*, New York, Modern Random House, Inc., 1993, p.81).

At first glance, it seems disturbing to consider the implication that, on one hand, the totality of human spirits represent a component of God, and yet, human spirits need to achieve perfection through the trials of the material world. Together, these claims seem to suggest an assertion that God is somehow less than perfect, but such is not the case. Our difficulty arises from our linear assumptions. We envision a situation in which:

First there is God alone—

Later God creates this thing called the soul of man, out of God's own substance—

Then God decides that this spirit, though holy, is not perfect—because it hasn't had an opportunity to exercise its love as an expression of free will, so—

Next, God creates the material world, and—

Finally, the spirit is perfected, and God is complete.

But we've already seen that time is just one aspect of the material universe, and we have taken, on faith, that the universe is itself internal to God. God is not caught up in the relentless current that affects all corporeal creatures. Thus, God does not *become*, because the whole concept of becoming is contingent upon changing over time. God, however, is external to time, and hence, unchanging. The path of perfecting, the tempering of that aspect of God's being, is an integral part of His perfection. It is how that perfection is manifest, rather than a *process* through which spirits pass on some path to improvement.

The Common Nature

Lossky states that "the Holy Spirit grants to each person created in the image of God the possibility of fulfilling the likeness of common nature."[5] This raises the question of precisely what is meant by "created in the image of God"? Clearly, the statement refers to the soul. Could anything that is the image of God not share in divine attributes? Hence, we assert yet again that the Universal Spirit is the unity of souls, and the Common Nature what we have referred to as the Christ-consciousness.

[5] Vladimir Lossky, *The Mystical Theology of the Eastern Church*, St Vladimirs Seminary Press, 1957.

John 16:13 says "Where the Spirit of truth comes, it will guide you into all the truth." Here again, we are to understand that this is where we must turn for truthful guidance in order to navigate through the decision points of life. But what exactly do we mean by decision points?

Drivers and Decision Points

There are only two drivers that influence us at each decision point, regardless of how significant or trivial it may seem: fear or love/compassion. Abstractly speaking, these correspond to the material, and the spiritual. (This is easy to see when you consider that compassion has no real survival value in a purely physical evolutionary process, therefore, it must be associated with the 'evolution' of the spirit.)

Every decision chooses a course that furthers the interest of either one or the other. Since the body is the seat of the five senses, and hence, the rational mind, its messages are generally louder and easier to hear. One must listen carefully, actively, in order to hear the more subtle voice of the spirit. The more you listen, however, the more attuned you will become to its frequency, so that eventually it will become the more dominant, more natural voice.

Thinking about it in this way makes it easier to understand the difference between the average person and, for example, a Gandhi or a Mother Theresa. We can understand instinctively how the inner voice of the saintly must have guided them in the ways it did, and can deduce the kinds of messages they must necessarily have heard and followed.[6] Logic, conversely, dominates the inner voice with a rather different set of messages. Since logic is concerned only with the physical, it compares working in the ghettos, among lepers, often in physical deprivation, against a life of luxury, expensive toys, fast cars, and so forth. Invariably, it tries to tip the scales in favor of the material.[7] If you truly listen for the voice of your soul, however, it will inevitably tell you that the expensive car you

[6] Once again, when we use such examples, it must be understood that it is merely for the sake of convenience. Every variation is available for every life, but for meaningful communication, we can only draw upon those which are accessible to the world-lines of reader and author.

[7] A life of total materialism is certainly a valid experience, for how better to understand the limitations of material satisfaction than to actually experience it?

coveted really didn't make you significantly happier than a more modest vehicle. Similarly, it will remind you of the great joy you received at having worked an act of kindness, regardless of how small.

For example, think back to the last time you exhibited an act of courtesy to someone else on the road, and they waved at you to express their appreciation. Didn't you feel elated? In fact, didn't the feeling of joy you experienced far exceed the act itself (at least in rational terms)? There's an important lesson in this: your discerning mind tends to devalue these joys because they add no benefit in terms it understands. It tries to free itself from memories of experiences that do not add value in the material premises by which it is served.

The physical body yearns for material wealth and for external power over others, which is its way of dealing with fears associated with doing without, or loss of external control. Let's go back to the last time you were behind the wheel, but instead of slowing down to let someone slip in, you exercised your external power by speeding up to prevent them from merging. Maybe they reacted by extending a middle finger, mouthing a pantomime imprecation, or some similar gesture. How did you feel? Probably physically exhilarated, as if you had prevailed in some kind of contest. Your heart raced, your jaw clenched, and you continued into your day primed for the next battle. How does this sensation compare with the earlier feeling of having done an act of kindness?

This is the crux of the story of Adam and Eve. Our rational mind often asks: "So what's so bad about eating from the tree of knowledge? Isn't knowledge a *good* thing?"

The lesson is difficult to assimilate because the rational mind values only material things and their associated knowledge, rather than spiritual wisdom. The willful choice to favor discriminating knowledge over spiritual truth is an act that feeds the ego. Did Adam and Eve, metaphorically speaking, already know this? Of course they did, because this was the very sin (ego) for which Lucifer had been expunged from heaven!

Each time you make a decision, you basically repeat the thought process followed by Adam and Eve on that fateful occasion—whether to navigate toward spiritual paradise or material satisfaction.

Conscious Navigation

Earlier, we used the metaphor of a cosmic scrabble game to illustrate the concept of conscious navigation through multiple-world decision points. In practice, the problem is that most people 'play the game' thoughtlessly. They simply grab random letters, so to speak, without a clear strategy. What is the object of 'the game'? Obviously, the objective is to learn the lessons that life offers us, recognizing the illusions of the material world for what they are, and to seek the perfection through which we will ultimately be rejoined with the Spirit of All. The means by which this is accomplished is the establishment of a full-time channel with your higher self. (In the Hermetic Order of the Golden Dawn, this is referred to as attaining the conversation with your Holy Guardian Angel.)

To some, this thought in itself may be a bit unnerving. Is the 'me' lost in this process of ascension?

To answer this question, we must understand that there are two parts that vie for the title of 'me'. One is the part of the personality that is seated in the brain. This part is the result of millions of years of physical evolution. It is the part that concerns itself with corporeal survival, material comfort—all of which are of little concern to the spirit. This part is itself eternal, being an active component of every discrete moment that makes up the fabric of eternity. It is, however, distinct from that other thing that experiences the life through its myriad of forking paths.

The spirit, or soul, (including that component that experiences physical reality through the organ of the brain), on the other hand, is the part of the individual that participates in the Union of Spirits, and hence in the perfection of The Whole.

Does this imply that the physical component of personality is evil?

We'll investigate the notion of evil in some depth later, but for now, let's simply speculate that since everything is part of God, nothing is inherently evil *per se*. Moreover, it's also highly likely that the topology of the brain represents the interface through which the spirit interacts with the physical realm, so we shouldn't be too quick to consider it in a purely material context.

From this perspective, we can say that the personality is not actually evil, but rather, that certain actions and events tend to impede the refinement of the spirit. On the other hand, to the extent that the personality is the seat of the ego, and preservation of the ego is the act of pride, it might be argued that pride is the only true sin, because it impedes the spirit. Looking at it differently, pride corresponds to any decision that exalts individuality when uniqueness solely results in divergence from perfection.

If we get into the habit of looking at things in this way, the channel with the higher self is established, and conscious navigation is the result.

Examine the threads you've spun since childhood in your current life-line. Evaluate the paths you have chosen, the realities you have elected to experience, and the ensuing results that have brought you to where you are in the *now*. You will begin to see your future choices with greater clarity. At each decision point you will see the different worlds you may choose to experience. Ask yourself which one you want to live in; which you wish to instantiate.

As you start to 'navigate consciously' you will find more and more coincidences beginning to favor your progress. Fewer impediments will hit you from left field. You will experience what you would once have interpreted as a change in luck. If you examine the twists and turns of your life in this light, it will become clearer how events fit into the way you have navigated.

The Mechanisms of Karma

Many Western religions have great problems with the notion of Karma. This aversion is difficult to understand from any perspective. Life is a tough test, nobody will deny it. Does it make sense to imagine that a merciful God would give you just one chance—no makeup exams—and if you flunk, off to hell you go for all eternity? If that were the case, one would have to conclude that He doesn't like us very much, and doesn't *want* us to succeed.

On the other hand, some mainstream denominations have eased this requirement by introducing the notion of purgatory. Purgatory is sort of a halfway house—something like hell, but a shorter sentence. If you screw up in life, you suffer for your transgressions, but just for a while. Obviously, this is the same mindset that thinks prisons should not aspire to rehabilitate, but merely to punish. To me, this doesn't sound like the philosophy of an enlightened God either, nor does it seem consistent with the message of Jesus.

In traditional terms, the notion of karma basically holds that what is not learned in one life is carried through to another, along with new required lessons. Rather than burning for our sins as a result of divine retribution, reincarnation describes a *voluntary* process through which we aspire to attain union with God.

Drawing on our earlier discussions, it would appear that one of the primary tools of karma is our material ignorance of anything outside of our monotonic self-awareness. Obviously, if we could see the 'big picture', then there's no way we would ever make a wrong choice. (Perhaps this is what differentiated the great saviors from others—while they *were* tempted, their vaster perspective precluded ever making the wrong choice. Hence, the *probabilities* associated with *their* alternate life-lines are so minute as to be considered negligible.)

Many, if not most, mainstream Western religions have some concept of punishment for transgressions, usually in the form of eternal damnation in a dismal region of everlasting agony. To the ancient Greeks, it was Hades, to others it's the fiery pit of hell. In my early education at a Catholic parochial school, a nun once provided the most memorable—and to me, most convincing—description of hell. She explained that hell is being deprived of the direct communion with God—the so-called Beatific Vision. It took me years to make the connection that this is also a description of life on the physical plane.

Some have consequently associated the 'process' of incarnation as equivalent to purgatory. Here we are honed and perfected, in an environment within which we are deprived of the Beatific Vision, for 'as long' as it takes for us to learn our lessons. In this view, karma is a causative process: You come to earth, you screw up in a variety of ways, so 'next' life, you are confronted with situations designed to correct these 'prior' shortcomings.

Within the model presented in the preceding chapters, it's easy to see why this model is inherently flawed. We are here to absorb these lessons from every perspective, for they represent the *experience* for which we are ultimately here. We've come to *feel* stuff, good, bad, boring, gleeful, agonizing, morbid, horrifying, elating, inspired, hopeless, and every variation in between. Could you imagine a god so dull and unimaginative that he would go to the trouble of creating a material universe, just for an endless variation on blissful experiences? That, my friend, would truly be hell!

Part of our mortal fear of what we perceive as ignominious fates is predicated upon the apprehension that this may be our only chance at material life. If we *know* that we get to experience precisely what we wish, as many times as we care to try, why *not* experience the martyr's death in the coliseum, or the hopelessness of the drug addict, or death by some erotic torture, or the feelings of love and dependence described in *Tuesdays with Morrie*.[8] Can we not see value in the life of Physicist Stephan Hawking, also dying of ALS?

Hence, Karma cannot be viewed as a punishment, but rather, the selection of lives we have made for ourselves, and their attendant interconnections.

The crucial concept here is that we are empowered to guide our own destiny. There is nothing that can be done to us through external means that can or would impede our spiritual progress. But, as in all of the more mundane aspects of life, empowerment coincides with responsibility. This is one reason why it's so difficult for us to accept our power—because doing so means we have nobody else to blame if we fail to utilize it effectively.

Most of us have an unfortunate tendency to interpret every impulse or insight as if it originated in our own individual material mind, but this is not necessarily the case. By recognizing, or at least admitting to the possibility, that such insights might be external in origin, we may tend to give them greater credence. In truth, they are not really external to our selves, but only to our mundane, material selves. Such inspiration comes from that part of ourselves that possesses a broader scope of vision—the part that represents God within us.

Why, we may wonder, does this part of us not simply step in and tell us more clearly what it wants us to do? The answer is that doing so would defeat the purpose of life! Part of the material experience is the limitation of perception. The paradox is that the more attuned you become, the more you tend to make ethical, moral, compassionate choices, and the more the inner voice can help you. The 'trick' then is to begin to value your intuition more highly.

[8] 1997, Mitch Albom, *Tuesdays with Morrie*, New York, Doubleday—The book describes the experience of a man dying from Amyotrophic lateral sclerosis (ALS), commonly known as Lou Gehrig's disease.

In the end, karma is a simple matter of how closely we align our material decision-making process with the predilections of the spirit. By so doing, your material experiences will correspond with your own best interests. The alternative is to be relegated to one of the alternate life-lines intended to 'teach you a lesson'—the hard way.

Before tying together the philosophical implications that can be drawn from the models hitherto propounded, let's examine some of the concrete mechanisms through which karma is developed and controlled.

Will and Choice

Will is synonymous with intent. Not every conscious act of will results in action, but even unactioned intent affects us. Thus, the first step in controlling your destiny is learning to control the will. This, once again, entails looking carefully at all eventualities and understanding what the whole picture looks like. Do you will to be the version of Ted Bundy who dies in Florida's electric chair, the eyes of the world upon you? This is a valid experience for some spirit, but do you wish to be the material vehicle through which this act is played out for eternity? For some, this might be hell, perhaps not for others.

The core of the spirit's evolutionary process, as well as the fulcrum of the quantum event structure, is choice. Choosing responsibly entails traveling down the probability paths that will unfold as a result of each possible option, and choosing one based on the perceived consequences. There is no judgment here, for all of the variations are here to be experienced by the spirit. The question is, which material path do you wish to support? Is it your personal version of Jeffery Dahmer? Mother Theresa? Or Larry Flynt? You take your choice and live with the result—forever.

All too often, people choose unconsciously, at random, and hence, their development is much slower than it has to be because they are simply swept along by the tides of chance. In the absence of some sense of navigation, we are like the rudderless ship, buffeted between Scylla and Charybdis. Choosing consciously takes effort, but is this effort greater than dealing with the results of random, and often self-destructive choices? Moreover, like anything else, conscious choice becomes easier and more natural with practice.

The game of chess provides an apt analogy. Beginning players (and computers) play the game by attempting to traverse every possible combination of moves in the search for an optimum strategy. More advanced players are able to rely on heuristics to prune the decision tree, effectively eliminating entire branches that will obviously lead to undesirable outcomes. Similarly, there are some paths that we *know* will lead to undesirable conclusions, without considering the matter through to its most minute detail. We need only a glance to tell us that *this* path leads us to a world that we really don't care to enter.

Temptation

Used positively, temptation is an opportunity—a chance to foresee the implications of what would ultimately turn out to be a negative karmic experience if actually carried out. It is the means through which our minds contemplate one available choice, and hopefully, reject that choice as inappropriate for spiritual growth. In this sense, you can think of it as a chance to learn one of life's lessons in something other than the hard way. From another perspective, it changes the probabilities of one life-line with respect to the others.

Are these choices easy? Of course not! There would be no increase in strength if the choices associated with overcoming temptation did not require sacrifice and discipline, but like anything else, making the right choices gets easier with practice.

Conceptual Difficulties

Many people brought up in mainstream Western denominations and inundated with Western rationality find conceptual difficulties with the notions of karma and rebirth. The most perplexing of these arise, once again, from our limited concept of time. Our inclination is to think in a strictly linear fashion. Thus, we think in terms that begin in prehistoric times, envisioning, for example, a soul working its way forward through stone-age incarnations, perhaps on to ancient Egyptian lives, maybe a medieval European incarnation or two, and so on through modern times and into the future. This, of course, leads to numerous difficulties. How do we account for the increase in population? Are new souls being created all the time? (Since it's likely that the spirit exists outside of time, this notion is untenable.)

What about the perception that our modern times often seem to embody a *decrease* in spirituality?

By now, of course, the reader should understand that the constraints of time are merely an artifact of our three-dimensional, material perspective, and are not at all absolute.

Try to envision our spiritual selves as we consider the lessons learned in our "most recent" life.

Where are we?

We are outside all of the physical universes, outside of time. What do we see? We see before us all of these opportunities to learn, experience, expand. They are stretched out like a vast intricate field. They branch and rebranch in a level of complexity that would be daunting to the physical mind. However, unencumbered by time or physical restrictions, we can study each in its entirety. We can see Plato over here, Julius Caesar over there, perhaps Bill Clinton over yonder. The fact that we may have just finished a stretch that began somewhere in the twentieth century and ran sometime into the twenty-first is really irrelevant in terms of our choices. We can go backwards or forward, as required/desired. We are also free to sample lives at our whim, regardless of what other spirits may have partaken of them. After all, the details of the life are already preordained, in all variations. There is nothing we can do to change that. We merely take a sip, then move on.

One result of freeing our thinking from temporal constraints is the possibility (as we've alluded to before) that the same spirit could, conceivably, occupy more than one material body in the same time period. In other words you (that is, the spiritual you) may be both the perpetrator and the victim of your own abuses; both the benefactor and the recipient of your own acts of charity. Thus, the admonition to "do unto others as you would have them do unto you" takes on an entirely more immediate aspect.

The point, however, remains the same. Regardless of how you look at things; whether or not you even accept the possibility of reincarnation, or maintain that we only get a single chance, one conclusion is inescapable: there are always persuasive reasons to do things in the best way possible.

Philosophical Implications

Having examined most of the relevant areas of mainstream (and not so mainstream) physics, we might well ask ourselves what is the purpose of probing deeper and deeper into the innards of matter, and outward toward the unfathomable reaches of space? Is it to pervert this knowledge toward the development of more cataclysmic weapons, as many have done, or even toward more benign technologies? Or should the purpose of these investigations be to better understand our relationship to the Universal?

What we've seen in the preceding pages is that although these concepts can't *prove* the existence of God (as Spinoza attempted), what they *can* do is help us to start thinking along lines that will lead us toward a direct experience of The Universal, and make that experience easier to accept.

In the model we've thus far established, we've seen how, in Biblical terms, God creates the material world driven by a desire to *experience*. Experience what? Clearly, whatever He has created! *All* of it! The sweet and the bitter, the joy and the sorrow, the pleasurable and the agonizing. Otherwise, why create it?

From our simplistic viewpoint, constrained by the three-dimensional limitations of the material, we perceive this world as having matured over millions of years, and call this evolution. From a wider perspective, unconstrained by time, we can conceive of how the totality of existence is, in fact, instantaneous, or more properly, totally un-timeable.

From the perspective of the Spirit, the material world, including the vast expanse of time, represents a wonderful vessel full of experiences to be engaged. Just as a dream that seems to entail hours, or even days really represents an instantaneous firing of brain cells, some of which entail the experience of time, so too do lives represent an instantaneous absorption of experiences that appear to involve time only from within that experience. Hence, it is easy to understand how a spirit might reach out to fulfill itself with the experience of a horse, or a dog, or an ant, as well as a king, or a killer; it's not like the ant sipping from a fire hose, but rather, the spirit delicately sampling what that material life has to offer.

Good Versus Evil

Let's come at the issue of good versus evil from a slightly different angle.

A question frequently raised by so-called 'people of faith', is: "how can an all-good, all-loving God allow evil to exist?"

God represents the ultimate consciousness, that which creates not in the form of actions over time (for that would make God subordinate to time), but as an instantaneous act of volition. His purpose is to experience, and to do this He has divided Himself into discrete souls through which He interacts with material reality. You are one of those components, as am I.

So, with this in mind, we can ask ourselves: What kinds of things would have to be in this material world for spirits to experience? Well, we would expect to find love and joy, trials and tribulations, successes and failures, miracles bright and horrors so abysmal as to shatter the senses. Disease, catastrophes, war, famine, and injustice would also be intrinsic components of this environment. Thus, when people wonder why a good God would allow evil things to happen, it's because we really don't understand that these are also the very things we've come to experience.

Once again, our fear of suffering and death is predicated upon a tacit assumption that we only get one chance at life. The intense agony of a Martyr's death in the arena, naked to the crowd, devoured by lions, or boiled in oil, is problematic only if one assumes they're sacrificing their only pass through life. But if it is not, then this is just one of a rich array of physical, mental, and emotional encounters that constitute our wider experience.

Remember, you have all eternity to experience what life has to offer—do you really want to spend it blissfully sitting on a cloud strumming a harp? So the next time you're wrestling with the question of 'why bad things happen to good people', step back and recognize that this is precisely why they're here.

This is the real meaning of the Bible's Book of Job. From a simplistic perspective, it appears as if an innocent man is being shamelessly abused by his God, but this very despair of being righteous but afflicted is precisely the experience for which he has come. He simply can't recognize the fact while 'on the ride'.

Moreover, if the point of the exercise is to perfect the soul, then endless happiness and bliss would do very little to propel that perfection. Consider the following:

If strength of character and courage are to be developed, then do we not need an environment in which we must grapple with hardship and react to disappointments? If service to others is valued, then must there not be opportunities to encounter social inequality? If hope is to be developed, then wouldn't we need the opportunity to be confronted by insecurities and uncertainties? If faith is desired, then must we not find ourselves in predicaments where we know less than we can believe? If a love and respect for truth are to be developed, there must necessarily be a possibility for falsehood. If idealism is desired then we must struggle in an environment that stimulates us to reach for better things. If loyalty and devotion to duty are desirable, then we must carry on amidst possibilities of betrayal and desertion. If unselfishness (the spirit of self-denial) is valued, then we must live face to face with the incessant clamoring within the self for recognition and honor. And what would be pleasure if we did not live in a world where the alternatives of pain and suffering are experiential possibilities? How could we ever judge something good if we could not experience its opposite? [Adapted from *The Urantia Book.*[1]] In other words, it's all here for a reason; all here for us to experience and appreciate.

But this brings us back to the question: What is the nature of good and evil?

Well, clearly, since all possible outcomes are part of each life, and those are what we're here to experience, everyone with whom we interact is in fact our partner, there to help us achieve

[1] 1955, *The Urantia Book*, Chicago, The Urantia Foundation.

our goals for being there. We come to *experience*, and must pay homage to those who help us do so, even when they seem to be our tormentors.

The ancient Chinese symbol called the *T'ai-chi T'u*, better known as the 'yin-yang' (Figure 34) exemplifies this concept.

Figure 34

Some people, mainly the most orthodox of fundamentalist Christians, abhor this symbol, mistakenly interpreting it as exalting the dark side in combination with the good. The more enlightened recognize that it simply represents an acknowledgement that all things serve God's purpose, regardless of how they may appear from our limited human perspective. For the Western mind, this concept can be difficult to grasp, but even the most stolid fundamentalists should be familiar with Job 38-41. Here, God chastises Job for presuming to understand things that can only be perceived from the divine perspective. Can even the most literal-minded evangelicals believe that such a thing as 'evil' could exist if God did not will it so, and if it did not serve His purpose?

An example that comes to mind is the case of Armin Meiwes, and Bernd-Juergen Brandes, who together implemented the German Cannibalism case. For those who are unaware of this situation, Bernd-Juergen Brandes, a young computer programmer, contacted Armin Meiwes via an internet cannibalism news group, and expressed the desire to really be killed and eaten. The two men came to an agreement by which Armin Meiwes began by cutting off the penis of Bernd-Juergen Brandes, cooking it up and cutting it in half so the two men could eat it together. Armin Meiwes, then killed Bernd-Juergen Brandes, and ate approximately 20 kg. of his flesh over the next year or so, before being caught. He provided police with video recordings of the event. (Initially convicted only of manslaughter, Meiwes was retried and sentenced to life in prison.)

The point here is that, despite our normal reactions of revulsion, we must realize that these feelings are the result of societal conditioning, not some inherent apprehension of good and evil. There is no evil here. There are merely two spirits playing out one of a myriad of possible interactions and experiences. Part of that experience is Armin Meiwes's publicity, in which he must experience the world's eyes upon him, knowing that he has broken one of society's most sacred taboos, and spending the remainder of his life in jail.

Lucifer's Transgression

We must bear in mind that everything seems very different from in here—the material world—than it really is. Souls are honed by the interaction between love and adversity. God, knew there had to be an icon—the image of what a perfect man would be—this is The Christ, as instantiated in Jesus, Buddha, Mohammed, Shango, and many others. They are the focus. They provide the model that must be followed to complete the maze. They embody the spirit that represents the aspect of God inherent in God's model for humanity. They are what God had in mind when he created man. God also knew that there had to be an architect of adversity, an icon responsible for spurring the spirits on, sometimes through means that, from the purely material perspective, seem harsh or even evil.

The existence of evil is a mystery mankind has struggled with from the beginning. The thing is, we're all working to fulfill God's purpose. How could it be otherwise? Consider that in Job (1:6-12, 2:1-7, etc.) Lucifer comes before God as one of the "sons of God," and the suffering he inflicts upon Job is, when you get right down to it, *God's* will. Lucifer is part of God's plan—to even consider otherwise is absurd.

Does this mean that Lucifer is actually sinless? The answer, of course, is a definitive no. His, in fact, is the only true sin—the sin of ego, what some call pride. In a very real sense, 'Lucifer's work' is really the toughest of God's work on earth. It's the stuff that's got to be done, but it's also the spiritual equivalent of a suicide mission; a death sentence—an eternal death sentence. God needed this service, but serving him in this way entailed eternal separation, imprisonment in this material bubble, from which Lucifer might never again experience direct contact with the creator.

Right now, locked in our material bodies, we have no real concept of what that means. One day, however, we will experience that contact, that Beatific Vision, which Lucifer is forever denied. For a spirit to believe it could bear that awful loss, even though the loss is incurred in service to God, is the damning sin of unmitigated pride. Thus is Lucifer damned for eternity, as the spirit of the material world. Lucifer is that part of each of us that is embedded in the material moments of time.

Applying the Concepts

The question still to be answered is: how can we *apply* this model of the universe we have so laboriously established?

The main impetus in preparing this treatise has been to remove an impediment. No words can convey the essence of reality, but many concepts can effectively prevent us from opening ourselves to gnosis. Chief among these has been the mechanistic doctrines of scientific nihilism. By gathering together those teachings that support the *possibility* of spiritual content, we eradicate that blockage so the mind becomes more open to accept the vision of enlightenment.

Imagine yourself looking at an enormous film reel, which you create by selecting each frame from a multitudinous choice of options. This is the film you yourself produce—your life's work. Will it be a masterpiece? A work of depravity, or perhaps worse still, of inconsequence? It has been said that in the moments before death, one's life passes before their eyes. What will you see when you gaze back upon what may be your greatest act of creation? The script is up to you. The details of our lives, the events and our reactions to them, are, in a very real respect, the pictures we create to be viewed by eternity.

Even if we reject all conjectures, and accept only the most conservative facets of physical science, these conclusions are inescapable. Additional faith only serves to build an even stronger case for moral, ethical, spiritual choices in our lives. But how can the things we've learned be applied in support of such choices?

To begin with, it should now be clear that there is no incompatibility between provable scientific fact and the possibility that each of us might have conscious access to caring, compassionate help in the analysis of our possible choices and their probable consequences. We need not assume that these come from anything external to ourselves. Nevertheless, since most spiritual choices can be seen not to serve the purposes of base *physical* survival or evolution, we are compelled to accept that if these drivers *do* originate from within ourselves, then it is a part of ourselves that is finer than that associated with pure animal instinct. The benefit of these drivers must, therefore, relate to *spiritual* survival and evolution. This observation, combined with some of the Eastern concepts of enlightenment, suggest an answer. In order to access this spiritual assistance we must obviously be open to *receiving* such help. To reach this state we must remove impediments placed before us by our so-called rational mind.

For hundreds of years, secular science has labored to convince us of the absurdity of all things spiritual. While sneering at the religious or mystical, it has steadfastly maintained that the only valid faith was that placed in its own mythology, smugly unaware that this was all it really was. Now, reluctantly, science has been forced to acknowledge that it does not have all of the answers. In fact, the answers may be inherently unknowable via the approaches it has so long propounded. Science now acknowledges that consciousness is indeed intimately connected with apparent reality.[2] Still, they cannot bring themselves to admit to the concept of spirituality, since doing so will, in a very real sense, invalidate everything they have come to believe—their creed!

Are we saying then that all scientific facts are wrong? Obviously not—that, after all, is the point of this book. What we *are* suggesting is that maybe the mechanistic conclusions are not justified by those facts. What we present here is another set of conclusions predicated upon those same facts—conclusions that are far more justifiable and evocative than ones which state that the world around us is inherently meaningless, chaotic, and devoid of spirit.

[2] Here we refer to the illusion of reality that is available to our senses and instrumentation. The true underlying reality, as discussed elsewhere, is not affected by observation.

This, once again, is the purpose of striving to provide an intuitive feeling for the more spiritual interpretations proposed in this book. Ultimately, each individual must experience for him or herself the reality of the Universal Spirit. This may be interpreted as Eastern enlightenment, the personal relationship with Christ, conversation with the guardian angel, meeting your animal guide, or through any other of the innumerable cultural metaphors. Once you have first-hand experience, you will realize that these are all just different ways of verbalizing the same thing. How could it be otherwise? More importantly, only with the confidence conveyed by such experience can one truly make informed and reliable decisions about life's branches and destinations, and understand the true meaning of life's experiences and lessons. Only with this assurance can one request and accept the help of the Universal Spirit.

To the closed mind, intuitions are mere curiosities. To the open intellect they are answers (sometimes to questions from a part of you with greater perspective and comprehension). To the closed mind these intuitions come unpredictably. To the accepting spirit they will always come when requested, and can always be relied upon. Moreover, asking the question and seeking the answers will inevitably lead you toward their source. Of course, you may not immediately be capable of hearing the answers. They may not come in the form you expect. The more you get into the habit of asking, however, the more attuned you will become to the answers, and hence, the more reliable they will become. Like all other behaviors and talents, the ability to ask the right questions and recognize the answers is a learned behavior, honed by practice.

Back in my college days I participated in a psychology department experiment aimed at identifying the possibility of conscious control over circadian processes. Participants were asked to concentrate, just before falling asleep, on waking up at a certain time—fifteen minutes earlier than they would normally be awakened by their alarm clock. We maintained a journal to record the results. After a remarkably short time, the results revealed, most subjects were able to set their 'internal clock' to awaken them at precisely the desired time. (Moreover, most awoke feeling unusually refreshed, since their bodies had awakened them naturally, rather than as a result of an external disturbance). Many similar experiments have demonstrated that the human psyche is highly suggestible in the moments immediately prior to falling asleep.

With this in mind, here's an exercise to help you apply some of these concepts to improve your life: Each evening, as you fall asleep, think about starting the next day as a step toward greater spiritual awareness—creating a clearer idea of who you want to be. Make each day a resolution to take control of your destiny.

Eschatological Considerations

Eschatology refers to the branch of theology that deals with the end of the world and/or the end of humanity. In the months leading up to the millennial turn of the calendar, tension increased, and even those who professed disbelief sighed a secret breath of relief when the New Year day came and went without catastrophe. But was this reaction premature?

The Relevance of the Millennium

Armageddon, the final ascension, the end of the world—the visions of many prophets and the forecasts of numerous sects have predicted that some cataclysmic event would change the face of the earth, leaving only the righteous to survive. Nostradamus, Edgar Cayce, and the legends of various Native American tribes concur on the coming of a great upheaval. Why did the ancient Mayans choose December 21st, 2012 A.D., as the last entry in their 2300-year old, and otherwise highly accurate calendar?

What can we make of such premises, especially considering how the much-anticipated 'Y2K event' came and passed with hardly a glitch?

First of all, it's important to acknowledge that this refers to an event local to earth, not to the universe as a whole. As we've already seen, the universe really doesn't *become* anything—it simply *is*. Similar events may occur on other planets populated by spiritual beings, but the principle, as we will discuss below, would be the same.

It is an indisputable fact that many religious and mystical disciplines have anticipated major changes in spiritual and physical conditions on earth somewhere in the timeframe of the millennium. It would be a grave mistake, however, to assume this window has passed, just because our current calendar has declared this arbitrary boundary traversed. As many skeptics have observed, the precise date on our modern calendar would have been largely meaningless to prophets of ancient times. How could their predictions of a millennium cataclysm be accurate or even meaningful?

It should not be too difficult to comprehend why any relevant prophesies would be approximate in nature, considering the mechanisms we have already proposed.

To begin with, we can acknowledge that our particular convention for marking the passage of time is hardly universal (as our previous discussions of time and space have emphasized). Just because we say the millennium has passed doesn't necessarily mean that the 'window of opportunity' is now closed.

Our knowledge of the space-time fabric suggests that a likely mechanism for the operation of extra-sensory perception consists of an ability to detect powerful mental emanations through the relativistic 'folds' of the continuum. In the case of simple 'hauntings', what may actually be happening is that potent emotion penetrates the deceptively thin fabric of time, and is experienced at a distance (through time) by another. Thus, the 'ghost' is not intruding on a particular place in the 'now', but rather, the observer is peeking into another time through a breach created by the intensity of that spirit's passion. This being the case, we can surmise that the highly charged emotional reaction of the masses of earth's population to any major cataclysm occurring in the *vicinity* of the millennium would convey the notion of a millennium catastrophe. The concept of the millennium would come through, even though the incident might not necessarily occur at the hour, minute, and second of New Year's Day in the year two-thousand. If any major cataclysmic event were to occur within, say, ten or fifteen years before or after the millennium proper, the great emotional trauma of billions of souls would probably be focused on the image of the millennia. The predominant thought on people's minds would be something along the lines of: "Oh my God, the millennium doom prophesies were true!"

Consequently, the simultaneous scream at the end of the world represents a trauma that would be noticed by many psychics, and what they would hear is a cry that carried this image; not the calendar or the precise date, but the concept. Thus, the prophesies were stated in these terms, but were not necessarily accurate in detail. The important point, however, is that the great majority of prophesies, both ancient and modern, emphasize that for those who are spiritually prepared, this event will represent a beginning, not an end.

The next issue we need to reconcile with the model is this: since we have suggested a universe in which, essentially, everyone is saved, what becomes of those who do not survive the impending maelstrom?

Einstein's model of space-time, coupled with the predictions of Hugh Everett's Many-World interpretation of Quantum Theory, supports the possibility that there could conceivably be far fewer discrete souls than there are corporeal bodies walking the earth at any given time. This is because, as we have speculated earlier, the Universal Spirit, which represents both the aggregation of human souls, as well as an intrinsic aspect of the Person of God, exists outside of the constraints of time. This model, therefore, leaves open the possibility that an individual spirit's material incarnations are chosen on the basis of the lessons to be mastered, or experiences to be absorbed, free of temporal restrictions. Hence, multiple incarnations may take place within the same 'overlapping' time period.

In describing the ancient world of some 10.5 million years ago, well-known psychic Edgar Cayce stated that the number of souls on the Earth plane was 133,000,000.[1] Since time is not a progression, then isn't it conceivable that this is the actual number of discrete souls being perfected through the multitude of corporeal incarnations? As of November, 2006, the world population was estimated at approximately 6,668,334,000. This represents a ratio of just under 50 humans per soul. (Think of this the next time you're about to deal harshly with someone, since the chances are pretty good that it'll be you! Conversely, it may simply mean that 49 of every 50 humans is simply a soulless automaton) In either case, The implications for the notion of Armageddon are obvious. At some point in time, a 'weeding out' will occur, the results of which are that the earth's population will be reduced so that there will be one corporeal instance per spirit.

[1] Edgar Cayce, 5748-2, May 28, 1925.

How might such a selection take place?

Since God has been seen to work his miracles through the laws of the universe He has created, it is likely that it will happen through some natural selection process. We have already seen how those with a more spiritual orientation have, in the years leading up to the millennium, become increasingly aware that something is about to transpire, and have felt gravitated toward certain areas of the world. This awareness may be likened to a form of spiritual evolution, which, like its physical counterpart, may well result in the survival of the fittest.

Figure 35 (from the book, Conversations with Nostradamus [Vol. 2], page 296, by Dolores Cannon, included by permission of Dolores Cannon), shows one vision of the future map of the United States as it will exist following the great upheaval.

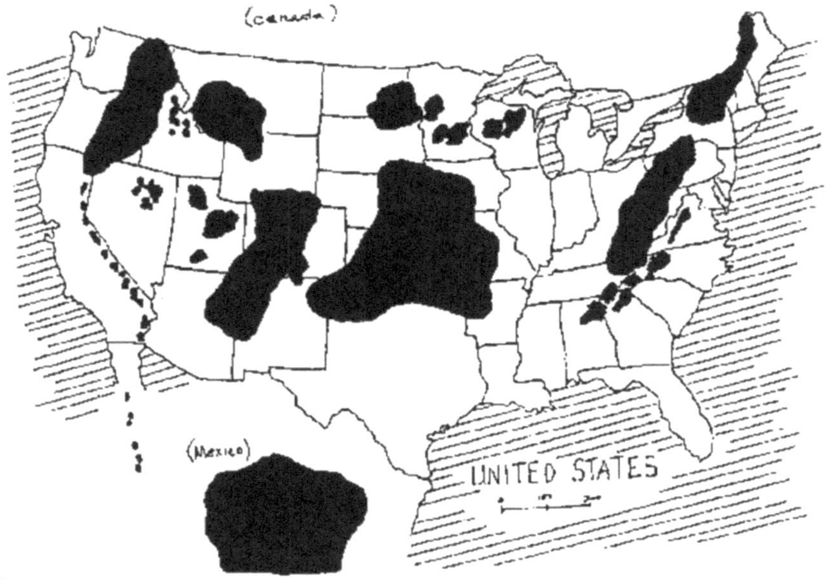

Figure 35

The black areas represent the remaining land that will be above water after the Earth shift. Figure 36 shows yet another prophetic vision of the future face of the American continent, this one from Ashton Pitre (printed by permission).

Figure 36

Numerous others have had visions which reveal surprisingly similar versions of our geological future.

Edgar Cayce referred to "those who may be given those directions as to how, where, the elect may be preserved for the replenishing again of the earth."[2] Thus, although many bodies would obviously be destroyed, all spirits will remain, each in the body through which they have been most successful in attaining spiritual maturation, as evidenced by their willingness to heed the call.

Materialism

The issue of heeding the call of the Spirit raises some interesting issues related to materialism.

Jesus compared the difficulty of a rich man entering heaven to passing a camel through the eye of a needle (Matthew 19:24). Similarly, in Luke 18:22, a rich man is told that he has to give away all he has in order to be saved. Does this mean that we need to become ascetics in order to attain salvation? Consider the story of Abraham (Genesis 22). God did not force Abraham to sacrifice his son Isaac, but Abraham did have to be *willing* to give up everything for his God. As Jesus says in Matthew 6:24, "no man can serve two masters."

[2] Edgar Cayce, 3653-1, January 7, 1944.

Many people in our present day, as in ancient times, make material concerns their master. Our present state of technology is such that people will soon be able to have enough to sustain themselves with very little effort, freeing up precious time for spiritual pursuits. Unfortunately, the vast majority (49 out of 50, I might speculate—the ratio of bodies that will be destroyed to souls that will survive) opt to spend every ounce of effort stockpiling material possessions at the expense of their souls. The few—the 1 in 50 who open themselves to the Spirit—will be willing to sacrifice material matters in order to go where the Spirit directs them. They, therefore, will be saved.

Of course, many modern prophets concur that coming earth changes may actually be modified by the preponderance of human thought patterns and attitudes. Most new-age predictions anticipate a time when human thought will exercise an instantaneous effect on the material world—a true age of magic. (Remember, the human brain might function like a transistor, allowing relatively weak patterns of intent to control far greater forces.) Strange as these predictions may be, the intimate linkage between human intent and quantum events may provide a mechanism through which such effects might be explained.

Ascension

We have already elaborated a model in which the totality of the human entity is represented as a combination of multiple components, only one small portion of which—the conscious material mind—is immediately involved in 'normal' day-to-day functioning within the physical world. Ascension refers to the state in which all portions of the self become integrated in a conscious way.

But why assume a continuation of the physical world at all? Wouldn't the final ascension render the physical world obsolete?

In one sense, we can look at the historical world—the world of wars, misery, and hatred as the ordeal of trial. From this perspective it would seem that the purified spirit would want to just fly up to heaven and be done with it. On the other hand, we can also see how the physical world, even the narrow cross section of which we are aware, is a thing of intricate beauty, not just something cooked up by God to torture us into purity. Perhaps, then, our ultimate fate is to enjoy the best of both worlds—to dwell in everlasting physical bliss, fully conversant with our spiritual wholeness.

Although the notion of eternal life in our physical bodies may seem more like fiction than a rational possibility, we've already seen the danger of extrapolating the certainty of previous experience as evidence of absolute truth. In this particular instance, experience suggests that all physical creatures grow old and eventually die, but is there a scientific reason why this should be so? Despite remarkable advances in physiological and biological research, scientists have yet to understand the mechanisms of aging, or the physical basis for death.

Conversely, we have seen how mind can influence physical processes in ways that are not at all obvious. In many cases these influences may actually appear contrary to intuitive logic. Examples include the bloodless piercings and firewalking exploits of yogis, the slowing of respiration and heartbeat by entranced practitioners, and the seemingly preternatural exploits of some Asian martial artists. Is it possible that, in a manner analogous to quantum events reacting to the observer's intent (i.e. we intend to see particles or waves so that is what we see), our own cells may react to our attitudes and beliefs? We might therefore speculate that we age at the cellular level because we believe very deeply that we *must* age and die.

The basis for this deeply seated conviction, of course, is that historically, this represents a pattern that's difficult to dispute. However, starting around the beginning of the twentieth century, virtually *every* scientific advance has represented completely unprecedented experience. Is it not reasonable to speculate that perhaps, one day, the mechanisms of aging will be understood and controlled?

From a metaphorical perspective, we might hypothesize that *this* was the meaning of the story of Adam and Eve eating from the tree of knowledge. The story is far more sensible if we understand that it was a tree not of true wisdom (imparted by the tree of life), but illusory knowledge of the physical world. Indeed, God *did* tell Adam and Eve quite explicitly that if they ate the fruit of that particular tree they would surely die, didn't He? But then the serpent—a creature closest to the earth—argued otherwise, and Adam and Eve elected to embrace physical knowledge, thereby condemning themselves to corporeal death. Did not Christ show us that man may transcend his fear and illusory limitations, even including death?

Most new-age prophets anticipate ascension as an event wherein our mind, body, and spirit harmoniously shift into some higher dimension or frequency. Many equate this coming resurrection with an impending change in the earth's polar orientation and magnetic reversal. There's little doubt that neural activity is essentially electrochemical in nature. Since we've already acknowledged that moving electrons induce a magnetic field, it doesn't require a great leap of faith to surmise that there could be some type of linkage between thought patterns and the earth's changing magnetic fields.

Some have suggested that the human body exhibits a kind of inductive resonance, responding to changes in the earth's fundamental vibratory states. Similar effects form the basis of integrated circuits, in which mutual induction results in numerous separate components exerting a synergetic effect on the whole. Likewise, there are persuasive indications that our own internal 'circuitry' does not operate independently of the surrounding fields of our environment.

As a scientist, my initial reaction to such speculations was to dismiss them as absurd. Upon deeper consideration, however, I began to wonder whether there might be some basis for these notions. In particular, recall our discussion of complex numbers, and the fact that the Schrödinger Equation of quantum mechanics describes a probability wave which has both real and complex (that is, existing in so-called 'imaginary' space) components. Since the Schrödinger equation describes the behavior of matter at its most basic level, might ascension be interpreted not as an increase in frequency, but rather, as a subtle transfer of full consciousness into this complex region, which is, we speculate, the domain of the spirit?

Regardless of how we understand or interpret these new-age predictions, one thing is manifestly clear: science provides an adequate basis for admitting to the possibilities.

As discussed earlier, it is likely that we exist as multi-dimensional beings that have infused portions of ourselves into the density of the material world. There are various beliefs and interpretations regarding this subject. The Bible, of course, reserves final ascension for the end of the world (with the exception of Jesus, of course); whereas many new-age belief systems hold that ascension can occur at any time, under appropriate circumstances. How do these conflicting views fit in with the scientific principles presented in this book?

If, as we have proposed, each spiritual entity is 'simultaneously' involved in multiple physical incarnations, then complete ascension would not be feasible until such circumstances as the less evolved incarnations have been culled out, and only the single, most worthy remains. This would seem to argue for the traditional biblical interpretation.

On the other hand, we've already speculated that certain individuals may have achieved this state at earlier periods. We might surmise that one of the distinguishing features of such individuals would be that, in some fashion, their lives represent singularities. This is entirely consistent with the concept we are now deriving about ascension; that is, part of their uniqueness involves a spiritual convergence on a specific corporeal life, so that ascension becomes possible.

In terms of new-age beliefs, the purported ascended masters would represent similar circumstances. In all self-honesty, the vast majority of us must acknowledge that our lives have strayed from perfection to a greater or lesser degree. On the other hand, it is manifestly obvious that ascension is probably not a one-step process. We can move toward perfection by degrees—first achieving an awareness of our spiritual surroundings, then working toward improved communication with our spiritual selves and guiding our daily lives accordingly.

Regardless of which interpretation we choose to embrace, one thing is clear: only through the effort of spiritual purification can we become candidates for ascension.

We earlier speculated that each of us represents a spiritual entity, only part of which is involved in the physical incarnation of the material being commonly thought of as the entirety of the self. We deduced that this larger spiritual entity might be involved in multiple physical incarnations, and, since the spiritual world exists external to the constraints of time, could be thought of as 'simultaneously' involved in the physical lives of its many physical manifestations. This entity, while individual, nonetheless participates in the essence of the Divine. In our stepwise logical development, we showed how such a concept is fully consistent with what we know of modern physics.

At this juncture, it should be obvious that most, if not all traditional spiritual beliefs and mystical practices can be supported by this logically structured universe, without the necessity of suspending one's credulity. Indeed, the intent of this detailed exposition has been to facilitate the pursuit of such practices with a mind unencumbered by doubt. In the next section we shall investigate a cross section of mystical concepts, showing specifically how they are supported by the model we have developed.

PART V:
SPIRITUAL CONCEPTS AND PARANORMAL PRACTICES

Many so-called paranormal, mystical, and spiritual events, concepts, and practices can be explained in terms of the world-model developed in this book. In this section we examine a cross section of these, analyzing them in the light of the models of reality established in the preceding chapters.

Although some of the topics covered in this chapter have been touched upon in various places throughout the text, it seemed like a good idea to gather them together here in one place.

The test of any theory or philosophy, as we've seen before, is the extent to which it is found to be consistent with observable phenomena or existing concepts. Clearly, since we began with scientific topics, we can expect the resulting model to be consonant with the tenets of physics, such as Relativity, Quantum Theory, Bell's Theorem, and so forth. Thus, the purpose of this section is to propose possible ways in which the model presented might be consistent with more esoteric concepts.

Astral Body

The idea that each human has an astral self, capable of separating from the physical body and journeying through time, space, and dimension, is an ancient one. Hindu writings describe eight *Siddhis* (magical powers) obtained through yoga, one of which is Astral Projection. Similar beliefs form an integral function in shamanic practices worldwide. Even the Catholic Church acknowledges an analog of this concept, which it refers to as *bilocation*, and uses to explain various accounts of saints being seen in two places at once.

In recent years the testimony of hundreds of people who have experienced 'clinical death' have been collected and published by medical researchers, the best known being Dr. Raymond Moody and Dr. Elisabeth Kubler-Ross. The experiences of these people who were revived and lived to tell what happened to them when they were 'dead' bear a striking similarity to those described in Theosophical literature describing astral travel and out-of-body techniques.

Although some schools of thought equate the Astral Body with the soul, it seems more likely that it represents some physical attribute distinct from both the physical body and the spirit. For example, astral body is frequently described as being the seat of desire, ego, and personality, which would suggest that it is something distinct from pure spirit. Other sources suggest that after death and separation from the physical body, the personality continues to exist for a period of time in the astral body. This also suggests that the Astral Body is some kind of interim state or interface between the physical and the spiritual.

This makes logical sense, when we consider what we've already discerned about the nature of the personality. We've already agreed (as we must) that *some* component of the personality is associated with the successive moments of the physical world, which are in turn eternal in nature. Even if not the ethereal soul, this abstract personality seems like it must entail something more than the mere electrochemical activity of the brain. Perhaps it represents the (mathematically) complex component of those quantum interactions that take place *in* the brain.

In any case, research in Kirian Photography, pioneered in the former Soviet Union, offers convincing evidence of the existence of tangible auras associated with living entities. The substance of these auras is, by some accounts, bioplasma; a material whose properties bear a striking resemblance to orgone energy, the life-force postulated by Wilhelm Reich.

Let's consider some other ways in which the concept of an Astral Body might be supported by the scientific principles examined in this book.

The second law of thermodynamics describes a measure of disorder known as *entropy*, and states that entropy in the universe must always increase (over time) in the absence of external organizational forces. The fact that physical life exists at all is an indication that there are external forces at work, which counteract the normal workings of entropy. (From another perspective, *some* energy must be extended to differentiate a human being from the mere puddle of goo that represents the raw materials of which he or she is composed.)

Clearly, mind represents a form or order, and this similarly suggests some underlying energy needed to overcome entropy.

Judging by the anecdotal evidence, it would appear that this thing, whatever its nature, has some kind of cohesiveness (i.e., is capable of functioning beyond the physical body).

Let's draw upon a computer analogy to better understand how these pieces might fit together. It's not too far-fetched to view the physical body as the computer hardware, with the astral body being the collected software that gives a particular system its individual capabilities. In an earlier chapter we speculated that, perhaps, multiple spirits might utilize the same body, each in its own particular world-line. This would be analogous to taking the same computer system and installing a completely new set of software components (perhaps a different operating system). We can also see how the software packages, as they exist on CD-ROMS or floppy disks, retain their capabilities even when not installed. Thus (in our loose analogy), the software files represent the Astral Body.

Where, then, is the actual soul or spirit?

Well, consider that any software package represents an implementation of some underlying principle. Before a word processor can be programmed, the designer must possess the abstract concept of text-processing procedures. Accounting principles exist quite independently of their instantiation in a bookkeeping package. Although there may be good and bad software implementations, we can easily see how the accounting principles themselves maintain a purity unaffected by specific instances. Hence, in this example the soul represents the underlying spiritual principle that exists quite independently of the behaviors, personality traits, and approaches through which that principle is exhibited in the context of physical reality. The astral body personifies those specific behavioral patterns as energy abstractions. Finally, the physical body is the mechanism through which those behaviors are played out in the material world.

(The body may, in fact, be an artifact of the interface between the astral body and the material world, in much the same spirit as we speculate that matter is merely an artifact of space-time curvature.)

As we have already discovered, all energy exists as patterns of space-time distortion embedded in a continuum that represents the reality of our physical universe. (Remember, the true spiritual plane is totally external to that physical reality.) All matter/energy, at the quantum level, also contains a complex component, which represents the linkage to the actual spiritual plane. Thus, we see how the Astral Body could represent those patterns of organized energies that correspond to our physical personalities. Since the material universe is finite, the Astral Body is not eternal, although its cohesiveness might well outlast the dissolution of the physical container (the body) in which it is developed. Nonetheless, it remains distinct from the spirit.

Remember, the material world probably consists of disconnected quantum instances or frames. In this view, the astral body might be that which provides continuity as it connects one particular path through the world-lines. Obviously, there would be distinct astral bodies for different paths through the matrix. The astral body of the Albert Einstein who remained a patent clerk throughout his life would be quite different from the one corresponding to the world-line we know.

Nonetheless, there remains a certain kinship between all of the astral bodies for a given individual. Perhaps in dream we are actually catching glimpses of these alternate versions of ourselves, often in perplexing jumbles and incomprehensible combinations.

As we shall see, this model also fits well into possible explanations for phenomena such as ghosts, apparitions, and so forth.

Astrology

Although many people will read their horoscope in the daily newspapers, most have difficulty taking seriously the notion that the position of the planets at the time of their birth could possibly affect their lives. This difficulty arises from assuming an inappropriate cause-effect relationship; that is, one in which the planetary alignments are the cause, and mortal fate is the effect. Nevertheless, it is a well-documented fact that many intelligent people, from presidents' wives to the heads of major stockbrokerage firms, consult astrological charts as part of their overall decision-support process. Considering that (in the latter case at least) we're talking about basically conservative, skeptical kinds of people, this suggests that the resulting information must have some empirical value.

A more plausible model emerges when we consider that the mechanism responsible may more closely resemble two items reflecting a single underlying influence, rather than one exercising a direct effect upon the other. We can easily accept, for example, how a strand of DNA extracted at random from any cell in a larger organism's body can, in principle at least, not only reveal every intrinsic detail about that organism, but also directly influence how that organism grows, reacts, and to some extent, behaves. Conversely, we can look at the macroscopic organism and make some pretty reliable predictions about its DNA. For example, we can easily tell that a certain person has genes for blue or brown eyes, light or dark complexion, pattern baldness, and so forth.

From yet another perspective, we know from Bell's theorem that reality must be nonlocal, that is, there are connections between seemingly distant events that take a shorter path than a straight line. While none of these considerations can *prove* that astrology works, they certainly serve to suggest mechanisms through which the position of matter in the universe, and the probabilities of events in our local vicinity of space, might be intimately connected, or might both reflect the influence of a more subtle aspect of the universe.

Similarly, Physicist David Bohm has suggested that the universe might be so radically interfolded that any piece will be a reflection of the whole. Couldn't ask for a more applicable theory than that!

In the final analysis, we know with near 100% certainty that the things we think of as representing reality are actually illusory reflections of something more fundamental. Hence, it shouldn't be too surprising if we learn that there are hitherto unsuspected connections at this more subtle level of reality.

Channeling

Channeling is a process through which information from a source external to the conscious mind is accessed and expressed. Virtually every culture has some history of institutionalized communication between living hosts and external sources of inspiration. Biblical prophesies, spiritual mediumship, and communication with extraterrestrial intelligences are all variations on this same basic phenomena (and probably just represent individual attempts to come to grips with what is always a rather startling experience).

Channeling has become a loaded word to many people. Basically, if you're a monotheist, when your own god talks to you it's divine inspiration, whereas when others hear the voices of other gods or spirits, it is necromancy, demonic possession, or a hoax.

Whether or not we retain the memory, most of us have reached the channeling state of consciousness at some point in our lives. It is a natural condition of being, which feels like inspiration, enthusiasm, and sheer clarity of thought. Afterward, you find yourself wondering, "where did *that* come from?" Thus, we can conclude that channeling is not an 'occult' state, but rather, a very natural skill.

So what about all these channeled 'entities', with strange names such as *Corfusalus,* or *Phylos the Thibetan,* or those claiming extraterrestrial origins?

Well, it's important to reiterate that consciousness is an intersection of our physical, experiential memories (probably resident in the astral body), with that greater perspective enjoyed by our soul. We receive inspiration in nonverbal images, then interpret it in terms of our physical memories. Thus, what Moses saw as a burning bush, another might interpret as a spirit guide, or the voice of the Pleidians (extraterrestrial spiritual entities).

Of course, this interpretation does not negate the possibility of communication with spiritual beings that are quite distinct from ourselves. In terms of channeling, a frequent concern is the influence of such possibly malevolent entities. Our traditional religions have given them many, many names, and we have all been taught to fear them.

In our discussion of the astral body, we described how this energetic component represents an aspect of the material personality that exists somewhat autonomously, and which could (in a temporal sense) outlast the dissolution of the physical body. It's not difficult to imagine what the astral bodies associated with, for example, Ted Bundy, Adolph Hitler, or Josef Mengele would be like. The traditional advice that applies here is: "by their fruits ye shall know them" (Matt 7:16). Even in your daily physical interactions, you usually know if someone you meet has intentions that are not in your best interests. Something, somewhere within you, whispers that something is amiss. The same applies to channeled sources.

Nor do we dismiss the possibility that there are souls that are inherently what we would think of as 'evil'. These are simply those aspects of God that are involved in experiencing the corresponding aspects of the material world. One can easily look at someone like Richard Speck, or Ted Bundy, and see precisely what they had incarnated to experience, including their ignominious ends.

Again, any judgment attached to their deeds are solely our own, and who are *we* to judge God? These are valid material experiences, and they are precisely what the spirit has come to absorb.

Such spirits might, in certain instances, wish to experience what it's like to 'possess' an unwilling host; but remember, the spirit of the host has incarnated precisely to understand the experience of being possessed!

In the majority of cases, channeling is overwhelmingly positive, and consists of establishing and broadening the bandwidth of that essential channel to the higher self.

Christianity (Orthodox)

I begin this section by emphasizing that I am referring here to a particular brand of the most orthodox Christianity that worships a version of Jesus who has been reduced to no more than a simple trickster, striding across lakes, turning water into wine, multiplying loaves and fishes, and other such slight-of-hand prestidigitation. These are the mindless followers of a flawed scripture, no more capable of rational analysis than a doorknob.

These people, furthermore, selectively ignore key scriptural elements from their own beloved gospels, such as when Jesus proclaimed that "he that beleiveth in me, the works that I do shall he do also; and greater works than these shall he do" (Luke 14:12); or Matthew 6:6, where Jesus says: "But when thou prayest, enter into thy closet, and when thou has shut the door, pray to they Father, which is in secret." Are these not the same hypocrites who constantly shriek their demands for public prayer, prayer in school, biblical effigies in court houses, and other public displays of their cockeyed faith?

From the concepts developed herein, it's easy to see that at the very least, the thoughts, actions, wisdom, suffering and death of *the version of Jesus with which we are familiar*, is forever embedded in the perpetual fabrics of eternity, just like everyone else. We need not wait for a second coming, for he is already here. Our task is to get in touch with this life, understand it, and either incorporate the lesson into our own life, or reject it as inapplicable to our individual mission.

We can go further and speculate that there are numerous versions of this life we know as Jesus, only one of which completely embodies what we have referred to as The Christ Consciousness; that is, in this one life, Jesus was completely in touch with his higher nature, and fully understood the mission he had come to complete. Might it be that *every* spirit ultimately experiences this manifold of life-lines, perhaps as the ultimate experience of their incarnations? After all, did he not say: “Nobody comes to the father but through me” (John 14:6)?

The most orthodox Christians hold many other notions that are a bit queer from any rational perspective. For example, they eagerly await a so-called ‘rapture’, when the just will just go floating away, and the evil will remain on earth to suffer. *They*, of course, are the righteous. Now where, exactly would these lucky winners be headed? To heaven? To a non-temporal dimension where their bodies, if they could survive, would be of absolutely no use to them? Where would they find a toilet in Heaven?

Of course, logic, reasoning, science, and consistency have little appeal to these people who have turned the life of this simple, humble, insightful teacher into the most overblown personality cult of all time.

I consider myself a Christian, to the extent that I believe Jesus existed, and that he provided a mode of life and thought that is intrinsically attractive to me. I try to emulate his example, spending my time writing books like this when, instead, I could be pulling down a three-figure hourly rate as a technology consultant. I try not to harbor hate, understanding that my enemies are merely working through their own karma, and mean no malice toward me; indeed, they are in integral aspect of my reaching my own goals, and as such, I must thank them as being worthy combatants.

I believe that perhaps Jesus *did* change water to wine, feed the multitudes on a few loaves and fishes, or perform any of the other so-called miracles of the New Testament, but I don’t believe these are in any way essential to the inner meaning of his life.

I am absolutely certain he has transcended death, to the extent that we all do by our very nature and by the nature of space-time itself. I suspect that some part of him, perhaps the astral body, appeared to others after his death—I believe others have done the same, and if anyone could do it, surely it would be a personality with the strength of Jesus.

One must also bear in mind that the current Canonical New Testament was a politically-motivated selection from a wide range of relevant material that was in circulation in the fifth century. Most noteworthy were the various so-called 'Gnostic Gospels'. Indeed it is manifestly clear that the growing church wanted to discourage the faithful from any belief that would reduce their dependency on the church. Hence, the Gnostic cornerstone of direct prehension of God was strictly taboo. One need only look toward the inclusion of the Book of Revelations to see that maintaining fear was a prime motivation on the part of the corrupt church. Clearly, nothing Jesus ever said remotely resembles this hodgepodge of horror stories intended to frighten believers into submission.

Getting back to our model, and drawing as well on much of the often contradictory Bible itself, we can see that there is absolutely no justification for the notion that we get only one shot at life, and after that, if we've screwed up, off to hell we go. There is no concordance with the idea that, at some time in the distant future, all bodies will be resurrected—they are, after all, eternal already. What are they going to do? Snatch them away from their places in history and line them up on the streets of America?

And, since all souls are inherently part of God, does it make sense that God would send himself to hell, separating himself from . . . God . . . forever? Remember what we said about an argument leading to an absurdity? Here's a prime example.

So, in conclusion, I'd have to say that there is nothing in our emergent philosophy that precludes belief in the historical Jesus, but that the hysterical creeds of the Orthodox Christians has absolutely no justification in our current view of reality.

Crystals

It's easy to appreciate how crystals must have impressed our progenitors as inherently mystical objects. Their brilliance, regularity, and color suggest an inherent source of order and perfection in a world all too often filled with chaos.

Crystals are solids formed by a regularly repeating internal arrangement of atoms and molecules, bounded by external plane faces. Crystal particles form a variety of geometrical shapes due to their internal compressions. In most solids, the arrangements of the building blocks (atoms and molecules) are random throughout the material. In crystals, however, collections of atoms called unit cells are repeated in exactly the same arrangement over and over throughout the entire material.

Crystals have numerous unique properties:

- As anyone who wears corrective lenses, or has gazed through binoculars or a telescope can attest, crystals can focus light waves.
- In 1730, Isaac Newton described the ability of a prism to separate white light into its constituent wavelengths.
- If an electrical current is applied to quartz, it vibrates at a very precise frequency. One result of this phenomenon is the ubiquitous quartz watch.
- Some crystals produce an electrical charge when compressed. This is known as the piezoelectric effect, and was discovered in 1880 by Pierre and Jacques Curie.

- Solar cells, semiconductors, and lasers are just a few recent developments based on the unique properties of certain crystals. Obviously, the electronics industry does not employ crystals because they are pretty, but because they order energy in unique ways.
- Kirlian photography has demonstrated that holding a quartz crystal intensifies the body's auras.

From the complexities of quantum mechanics to the simple beauty of Kirlian photography, the effects of crystals can scarcely be questioned.

Despite these incontrovertible phenomena, mainstream scientists steadfastly sneer at any suggestion that crystals might interact with the human organism in more subtle ways. Can we make a reasonable scientific case for arguing the possible metaphysical efficacy of crystals? I believe we can. Let's start by taking a closer look at a device that should, by now, be familiar to most readers: the laser.

Although they come in a variety of forms, the one most relevant to the current discussion utilizes a synthetic ruby crystal to "line up" photons so that they are perfectly interlocked (coherent), resulting in a powerful, highly concentrated beam of energy. The reason this is possible is due to the ability of atoms in some crystals to exist in a *metastable* state.

In most ordinary materials, when energy is absorbed (for example, through heating or exposure to light), the electrons of the material's atoms are bumped up to higher orbits. After a short amount of time they decay back to their ground states, emitting a quantity of energy equal to that which they absorbed. The decay may take place in a single step, or several smaller steps. For example, high-frequency ultraviolet radiation passes through the windshield of your car, and strikes the interior surfaces. The high-energy UV photons pump the electrons of the interior surfaces up multiple levels. The electrons decay in several smaller steps, emitting less energetic infrared (heat) photons at each step. Since the windshield is effectively opaque to infrared, the heat is trapped, and your car gets hot when parked in the sun.

The existence of a metastable state in some crystals changes this situation significantly. As before, external radiation pumps the electrons up to higher orbits. As they decay, however, they encounter the metastable orbit, and tend to gather there for a relatively long period of time, rather than decaying spontaneously. When they finally do decay, a cascade effect occurs, and the photons that are released are highly organized, and in perfect phase with one another. The result is the now familiar laser beam.

Yet another related subject, mentioned earlier, is that of *quasicrystals*. Quasicrystals possess an internal structure that can only be understood in terms of physical dimensions in excess of *five*. The observable crystalline structure of quasicrystals corresponds to a projection of the n-dimensional periodicity into three-dimensional space. The fact that this structure is required only for quasicrystals in no way prohibits the possibility that other crystals may also be three-dimensional projections of higher-dimensional objects. Hence, it is not unreasonable to speculate that crystals may interact with living entities through some higher-dimensional linkages.

As we have already seen, the universe is awash with random, undirected, chaotic energy. Life, of course, is an organized, energetic process, as is thought. For the purposes of the current discussion, we needn't concern ourselves with the source of energy responsible for this organization. Suffice it to say that this presence of order reveals that energy from some source has overcome entropy, resulting in stored potential energy. Similarly, it's not difficult to see how thought itself, being an ordered electrochemical process, must also be an energetic one.

We know with certainty that the mind is demonstrably an active participant in all observable quantum events. Hence, it is not at all far-fetched to hypothesize that the proven capabilities of crystals to focus and amplify energies might apply to *all* energies, including those of biological origin (after all, mass is mass, and energy is energy, regardless of its source). If crystals can organize energy simply by contacting it, it is not unreasonable to speculate that they might exert some reciprocal influence on our own subtle energies. As we discussed earlier, all mass and energy exist, at least within our local four-dimensional space-time matrix, as 'dimples' in the fabric of reality. From this perspective, it is not at all difficult to envision such subtle interactions.

Demons

Let's begin this particular discussion by setting aside, for the moment, any consideration of the supernatural, and asking ourselves whether we can identify any human entities that might reasonably be seen as the embodiment of what we think of as evil. Hitler? Perhaps, but I think better examples from that epoch might be some of the death camp participants who were more directly and personally involved in the well-documented atrocities; people like Josef Mengele, John Demjanjuk (Ivan the Terrible), and the like. What about people like Ted Bundy, Richard Speck, and John Wayne Gacy? These are people motivated not by the lust for power or even misguided political ideologies (like Timothy McVeigh), but rather, by a desire to inflict suffering purely for their own personal satisfaction. (But again, the victim has also incarnated for precisely that experience, as have their loved ones, everyone involved in the investigations, newspaper readers, and so forth.)

Considering the vast array of available examples, I think it's safe to conclude that we can identify certain intelligences that are inherently, unrepentantly, and irredeemably negative. Hence, without resorting to the supernatural, we can propose that, from some perspective, demons are the intelligences of those creatures whose actions we see as base in nature, and contrary to the direction indicated by the Christ-consciousness. We cannot, however, associate any judgment with such spirits, and indeed, at least some readers will *be* such spirits themselves. All *we* can do is ask ourselves whether *this* is why we think we are here, and whether it represents the direction in which we wish to travel.

Does this suggest that there are no such things as discreet supernatural intelligences, some of which may well be malignant in nature?

Not at all.

Recall how, in our discussion of the Astral Body, we posed the argument that the full complexity of the individual exceeds the boundaries of the physical, and that the Astral Body, which is distinct from the soul, represents patterns of thought and behavior—the body's programming, if you will—which exists beyond the temporal limitations of the physical shell. Since we've already identified the existence of inherently evil personalities, it stands to reason that the propensities of these individuals, freed from the constraints of the physical, would be indistinguishable from those which we traditionally attribute to demons.

This is not to say that there may not be other discreet intelligences present in the invisible realms of the material universe. Rather, it merely suggests that these are not *required* to explain our evil demons. If such entities indeed exist, than they are very likely to be explainable using the same scientific arguments.

Divination

Divination is one of the most popular spiritual/mystical practices. For the purposes of this discussion, I will define divination as any attempt to attain insight by mechanical means other than direct enlightenment. As such, specific divinatory tools represent aids or props that help the mind to overcome conscious inhibitions in order to exercise one of its natural capabilities. It follows, therefore, that divination is not ultimately concerned with mere fortune-telling, nor even with divining the spiritual causes of material events, but rather, with the development of the inner psychic faculty of intuition.

The object of divination is the construction of a mechanism through which this source of inspiration becomes accessible to the conscious mind. Thus, although these practices may result in more or less accurate answers to the specific questions posed, the intuitive development constitutes the most important aspect of the practice.

Albert Einstein is quoted as having said: "I never came upon any of my discoveries through the process of rational thinking." Nonetheless, the vast majority of mainstream scientists dismiss the possibility of knowledge originating from sources other than logical deduction.

In the preceding pages we have exhaustively explored the many ways in which the mind may both influence its surroundings, as well as act as recipient of impulses from beyond its immediate sensory range. In this sense, we see the various divinatory practices as enablers, facilitating the communication channel between our conscious mind and that aspect of ourselves whose vision spans greater vistas.

Considered in the light of the many-world model, divination takes on yet another perspective. If every alternative exists as a distinct and equally real world-line, then the information imparted by any divinatory practice could only represent an indication of likely outcomes based on the practitioner's current position, much like a roadmap.

Remember, Quantum Theory is all about probabilities, and different probabilities are associated with the multiple Everett world-lines. Hence, divination is a way to assess these probabilities, in a qualitative, intuitive manner, based on your current position and propensities.

This, is not at all inconsistent with traditional interpretations, which generally maintain that readings never indicate what *will* be, but only the possibilities. Once again, this model would tend to emphasize the importance of divination, not so much in terms of prediction, but rather, for enhancing and developing intuition and gnosis.

Dreaming

Throughout history, philosophers have wrestled with the notion of dreams. An often-repeated anecdote asks whether our so-called waking life might indeed be the dream, while what we think of as dreams are actually reality.

Dream interpretation, of course, has been a mainstay of divination in virtually all cultures, primitive and advanced. In the Bible, God repeatedly speaks to His prophets through dreams. As reported in Matthew 2:13, the importance of dreams cannot be overstated:

"An angel of the Lord appeared to Joseph in a dream. 'Get up,' he said, 'take the child and his mother and escape to Egypt. Stay there until I tell you, for Herod is going to search for the child to kill him'."

In Daniel, the king tells his astrologers: "This is what I have firmly decided: If you do not tell me what my dream was and interpret it, I will have you cut into pieces and your houses turned into piles of rubble. But if you tell me the dream and explain it, you will receive from me gifts and rewards and great honor. So tell me the dream and interpret it for me."

High stakes indeed!

Similarly, In Deuteronomy 13:5: "And that prophet or that dreamer of dreams shall be put to death, because he has spoken to turn you away from the Lord."

If we accept the multi-world theory, then it's easy to see how dreams may, in fact, be disjointed snatches of alternate realities. It may well be that the confusion that typifies many dreams is merely a matter of discipline. For example, practitioners of *lucid dreaming* develop an ability to gain awareness of their dream state, and consequently, to control the unfolding of events. Similarly, the distinction between chaotic dream and astral travel may be a continuum, based upon self-discipline.

Hence, it seems likely that for one disciplined enough to control them, dreams could indeed become a reliable channel for guidance, inspiration, and information.

Extra Sensory Perception (ESP)

Extra-sensory perception represents one of the most widely and reliably documented of the so-called paranormal phenomena. Psychology departments at numerous prestigious universities regularly sponsor research in this area. Police departments have employed it for solving difficult cases, and the governments of both the United States and the former Soviet Union have investigated its use for espionage.

ESP represents the passive branches of PSI,[1] and includes telepathy, clairvoyance, precognition, and mental control. Telepathy involves the gaining of information through the sensory input of another person, possibly separated by great distances. Clairvoyance, on the other hand, deals with gaining information from unknown channels, presumably in the spiritual realm. (See 'Channeling', above.) Precognition entails the acquisition of nondeductible information about future events. Finally, mental control, as the name implies, involves controlling the perceptions of others.

There have been convincing studies suggesting that, in at least some cases, extra-sensory perception or ESP appears to be real. This, in turn, infers some physical mechanism, at present ill-understood, through which these effects are instantiated.

[1] The term 'psi' was introduced to be a neutral term for a variety of 'paranormal' phenomena, free of the implicit assumptions that are contained within labels such as psychokinesis (movement by mind) or telepathy (feeling or perception at a distance).

In one of the better-documented demonstrations of ESP, Russian scientists employed a rather unique approach. Two test subjects, a sender and a receiver, were located in separate buildings some distance apart. The receiver was wired up to a brainwave monitor to record changes in his mental state. The sender was instructed to concentrate on images of doing violence to the receiver, but only when a light in front of him was turned on. Neither the sender nor the receiver were told the nature of the experiment. The light flashed on and off at irregular intervals (at least that's what the sender thought.)

Only afterward was it revealed that the flashing light represented a message in Morse Code. By examining the changes in the receiver's brain waves, the scientists were able to decode message perfectly. This resulted in one of the first truly quantitative demonstrations of a paranormal phenomenon. (The critical aspect of this experiment was that it really didn't matter whether the receiver was able to interpret the precise nature of the signals being sent to him. The scientists looked only at the spikes in the receiver's brainwaves, indicating when *something* was being sent to him.)

In the annals of Soviet research one of the most widely documented individuals with this ability was a fellow named Wolf Messing. On one occasion, his performance was witnessed by none other than Sigmund Freud and Albert Einstein. By way of a demonstration, Freud asked Messing to carry out the series of actions Freud was envisioning. Messing went to the bathroom, retrieved a pair of tweezers from the medicine cabinet, returned to Einstein, and apologetically plucked three hairs from his trademark mustache. This was, in fact, precisely the task that Freud had mentally set.

By all indications, ESP appears to be an energy phenomenon. (In so far as these are objective events, they can hardly be expected to be anything *but* energy processes.) The significance of such events is that if even *one* incident is proven to be true, then it must be taken as indicative of some domain of physical process beyond our current ken.

As we have already seen, Quantum Theory *proves* that the intent of the observer plays an active and intimate role in seemingly mechanical interactions. The work of John Bell has similarly proven that action-at-a-distance is demanded by the tenets of Quantum Theory. Finally, General Relativity proves that the space-time continuum is radically curved and folded. Together, these concepts suggest that so-called extra-sensory perception is indeed a phenomenon through which these intrinsic aspects of reality are exploited by certain individuals. Hence, these capabilities can be understood in terms of the world-view developed in the preceding chapters.

It seems highly likely that these are natural abilities inherent in the field interactions set up by the brain, but which are often unavailable due to inhibitions incurred from earliest childhood.

Ghosts and Apparitions

Most phenomena in this category are associated with emotional trauma of some sort. A violent or untimely death is a common thread. Even in so-called poltergeist experiences, the cause is frequently traced to the emotional influence of an overwrought teen.

Two scientific observations that seem to offer promise in explaining such phenomena are the possibility of a complexly (mathematically speaking) enfolded space-time continuum, and the influence of mind in underlying the most basic levels of physical phenomena. What makes a comprehensive definition particularly difficult is the wide variety of forms such manifestations seem to take; suggesting that the category actually consists of numerous distinct, but possibly related phenomena.

In many so-called hauntings, the apparition seems largely unaware of the presence of observers. This suggests that the specter is not, as commonly assumed, intruding on the present, but rather, because of the intensity of the subject's trauma, we are witnessing an event through a rift in the space-time continuum.

A related subject, discussed earlier, is that of the astral body. The astral body represents the patterns of thought that characterize a particular incarnation, freed from the constraints of the physical body. As such, it remains distinct from the soul or spirit, existing as a persistent shadow of the material entity. Thus, it retains much of the mundane desires, drives, and angst of the corporeal entity with which it was associated. Hence, it is likely that the component we perceive as a ghost may be identical with the astral body.

Magick

"Okay," you may be thinking, "I can buy into some of this other stuff having a scientific basis, but magick? Give me a break!"

Well, if you're thinking in terms of *Harry Potter*, or the lady in *Bewitched* twitching her nose and things floating through the air, then you're correct of course. On the other hand, theoreticians of magick suggest a model that is remarkably consonant with the world view developed in this book. It goes something like this:

We already know with near 100% certainty that the world around us is largely illusory. Ergo, this illusion must be a distortion *of* some underlying reality. According to magickal theory, reality is first manifest in this more primary plane (or realm of more extensive dimensionality), and only subsequently does its distorted reflection become apparent in our so-called material world. The purpose of the various magical practices, in this model, is to free up the mind so that it (or, more properly, the astral body) can cause changes in the astral plane. These changes are then reflected some time later in our material world. Obviously, such changes are rarely immediate, nor do they take the form of 'power beams' emanating from the business end of a magickal wand. Nevertheless, practitioners insist that such changes can be wrought as a result of their ritual conjuring.

Looking at it from this perspective, the prospect does not seem nearly so absurd, and is more or less consistent with what we have already learned about physical reality. We can also put a many-worlds spin on this. If, indeed, every possible eventuality is instantiated in the various world lines, magic emerges as no more than a method of navigating the practitioner's consciousness toward a world-line consistent with his or her desires.[1]

[1] Perhaps more cynically, we might suggest that it involves a set of experiences in which magic *seemed* to have manifested its desired outcome.

Microcosm/Macrocosm

The macrocosm-microcosm relationship was one of the central concepts of Renaissance natural philosophy. Put simply, the idea holds that the universe can be viewed as a hierarchically organized mechanism in which every part reproduces the whole. For example, the rules and forces shaping and governing the human body are the same as those shaping and governing the entire Universe.

This philosophy is echoed in the concept of the human soul being created "in the likeness" of God.

Physicist David Bohm, whom we mentioned earlier, has resurrected this concept by suggesting that reality may resemble a hologram. One of the salient aspects of holograms is that if any single part of a hologram is projected, the result is a reproduction of the entire image. Although lacking in some of the detail that would be visible from the complete hologram, the partial image is nonetheless complete. In Bohm's model, each constituent component of the real world may similarly reflect the overall structure of the whole. Thus, we see yet another situation in which physics tends to corroborate concepts originating in mysticism and alchemy.

Miracles

The dictionary defines the word miracle as: "an event that appears inexplicable by the laws of nature, held to be supernatural in origin or an act of God." In general, miracles may be interpreted as falling into three broad categories:

1. Miracles of timing—incidents that can, in and of themselves, be explained by the known laws of science, but which, due to auspicious timing, seem to defy the laws of probability.
2. Miracles of unknown mechanism—incidents which, although inexplicable in terms of current systems of knowledge, nonetheless appear to follow regular patterns that suggest some underlying principle.
3. Miracles of uniqueness—incidents that are totally extraordinary in terms of human history.

As an example of the first category, we might consider some recent interpretations of the parting of the Red Sea by Moses. Some scientists and historians have argued that a certain combination of tides, water flow anomalies, planetary alignments, and so forth could conceivably have resulted in an unusually low water level, followed by an abrupt inrush, at approximately the date suggested for the incident in biblical and historical accounts. In this case, even if the unusual physical circumstances can be explained by such a combination of natural phenomena, the timing alone would be nothing short of miraculous.

Examples of the second category might include incidents of faith healing, spectral appearances, mental telepathy, and so forth. Sufficient occurrences have been documented so as to suggest common features that indicate these may be the result of some aspect of nature that simply has not yet been identified and quantified.

As an example of the third classification we can hold up the story of Jesus walking on water, or feeding a multitude on a few loaves and fishes as being sufficiently unique in the history of human records to qualify.

One widely held view is that the "age of miracles" has passed. Some theologians argue that miracles were needed in earlier times, but now that we know of them through tradition and scriptures, we must rely on faith. (How convenient!) Secretly, many believe that these stories are apocryphal in their essence, misinterpreted from their origin, or mutated over the centuries into fables designed to emphasize powerful moral precepts.

It might reasonably be argued that the prevalence of miracles in biblical times was more a matter of people having been open to miracles. Today, we consider ourselves too sophisticated to believe in the preternatural, so we catalog those phenomena we cannot understand or identify as perceptional aberrations or simple misinterpretations, and move on with our lives. Thus, in the final analysis, miracles are all around us, if only we can open our eyes and acknowledge them.

Numerology

Numerology dates back to the ancient Hebrew, in which each letter also possessed a numerical value. Numerology (known as Gematria) consisted of the calculation of the numerical equivalence of letters, words, or phrases, in order to gain insight into the interrelation of divergent concepts. The assumption behind this technique was that numerical equivalencies were not coincidental. Thus, since the world was created through God's 'Word', each letter should represent some aspect of the creative force.

The underlying concept behind numerology is that all things have a unique vibration. For example, what we see as the color red is actually a vibration that is the same every time we see that color red. In numerology, the vibration of red is translated into numbers representing a light frequency, and that same number may also represent the frequency of a particular musical note. Thus, numbers are treated as a kind of shorthand for a variety of physical vibrations. Ultimately, mystical tradition understands the act of creation in terms of vibration. So, the words of Genesis: "In the beginning was the Word" are understood as "in the beginning was vibration."[1]

[1] The staunch atheist might argue that the beginning was merely a result of random quantum fluctuations in the primordial singularity. To this we must respond by asking: "what caused these 'fluctuations'?" Don't forget, the prevailing view holds that time and space came into existence at the moment of creation. Causality is inherently time-linked, so it really makes no sense to ask what 'caused' these fluctuations, unless it was some thing outside of time/space.

Since, as we've already seen, Quantum Theory treats all matter and energy in terms of the quantum wave function (waves are identical with vibrations), we can see how numerology might ultimately have some basis in physics.

Moreover, the notion that mathematical expressions, and numbers themselves, might have some kind of transcendent reality should not strike us as too farfetched. In the final analysis, mathematics might *be* the ultimate reality.

Prophesy

Prophesy has represented the most visible channel to the Universal Mind since the beginning of time. From shamans of the most primitive stone-age cultures, to the oracles of ancient Rome, and down through the present, even skeptics treat these possible conduits with a degree of reverence. Edgar Cayce, whose prophesies I quote at appropriate points in this text, is one of the better known modern-day prophets, along with Jean Dixon, and a handful of others. Ashton Pitre and Dolores Cannon have both presented prophesies, including detailed maps, illustrating the appearance of the world's land masses following anticipated cataclysmic earth upheavals.

Although it would clearly be naive to accept every crackpot prophesy without some degree of scrutiny, it would be equally unreasonable to dismiss such predictions without open-minded consideration. Most importantly, we must remain open to our own inner voices, and not minimize our own intuitive channels as possible sources of guidance, for ourselves or for others. Remember, such insights are frequently messages from that aspect of ourselves which enjoys a far broader scope of vision.

In my own case, the messages were startling as ice water. I was living in Massachusetts, and my mother was visiting for Thanksgiving. It was early morning. My wife and children were still asleep, and my mother and I were in the kitchen, drinking coffee and chatting. I remember looking at her and marveling at how gray her hair had gotten. Suddenly, I found myself . . . outside. Not outside of the house, but outside of *everything*. I could actually *see* time spread out before me as I have sought to describe throughout this book. In an instant, all of the philosophical conundrums seemed elementary, transparent. When I 'returned', my mother was still talking, and hadn't noticed a thing. It took me a moment to realize that it had all happened in a split second.

My initial feeling was a conviction that *this* was what I had been put on earth to do: to report what I had witnessed in a way that it would be understandable, and beneficial, to as many people as possible. Nonetheless, it took me ten years to even *start* writing the book you now read! Why? Well, to start with, I immediately started making excuses. I was just being silly, I tried to tell myself. I had never written a book—how would I even start? I wasn't a philosopher, I was a computer scientist—who would listen to me? Moreover, what would writing lunatic-fringe new-age gobbledygook do to my professional career? In fact, I could find *dozens* of reasons not to start.

Then, without going into gruesome details, things started going *wrong* in my life. During this time I tried to hedge my bets by writing a succession of fictional pieces that sought to expound this peculiar vision of reality I had experienced, without exposing myself to professional criticism. Still, my life seemed to remain stubbornly out of kilter.

I guess a big part of my reticence had to do with the whole concept of *prophesy* itself. It isn't easy to start thinking of yourself in those kinds of terms, much less to consider presenting yourself as such. It just seemed so . . . pretentious. Ultimately, though, I reached precisely the opposite conclusion. It isn't so much a question of me (that is, the goofy material me) being a source of such insightful observations, but rather, that these insights to which the mundane me could never lay claim, are transmitted with me as no more than amanuensis.

Finally, I bit the bullet and wrote the book you now have before you. Shortly thereafter, things started falling to place once again. I guess the story of Jonah is probably the best analogy. If the Big Boss drafts you, you'd better show up for basic training.

From what we have already come to understand about the nature of time, it is not unreasonable that certain individuals might somehow be capable of predicting future events. We've already seen that, according to Bell's Theorem, Quantum Theory demands nonlocality; that is, the notion that events which are apparently separated in time and space can be intimately connected in ways we are unable to fathom.

Just because we pose a scientific explanation, however, it would be a mistake to think that this mechanism is too complex, too intricate, or too circuitous to represent the channel through which The Universal elects to speak. In fact, we've repeatedly seen that the complexities of our scientific universe are inevitably the channels favored for any miracle.

Psychokinesis

Psychokinesis refers to the PSI phenomena wherein a person appears to directly affect their environment. Just as ESP is analogous to sensory functioning, psychokinesis is the PSI equivalent of motor functioning. Traditionally, Psychokinesis has been divided into two categories:

Micro PK applies to cases in which instrumentation and/or statistical analysis is required to determine if there is an effect. An example would be the ability to influence microelectronic devices.

Macro PK refers to situations in which naked-eye observation can perceive the effect. Examples include poltergeist activity, table tipping, and of course, the famous spoon-bending exhibitions.

Obviously, the first category is not only explained, but demanded by the basic tenets of Quantum Theory. The intent of the observer is intimately involved in all physical phenomena at the micro level. Consequently, the only question that remains is one of scale. Can the mind, under the proper circumstances, exert an influence over macroscopic objects?

A widely witnessed and recorded example of psychokinesis was the Russian woman named Nelya Mikhailova. Surrounded by cameras, enclosed in a metal cage to shield from any external electromagnetic influences, Nelya was repeatedly capable moving a single identified matchstick from a pile of identical sticks, all strewn under an inverted glass bowl. When performing these feats, Nalya's pulse was often raised as high as 250 beats per minute. Interestingly, it was reported that her PK abilities diminished during periods of stormy weather. This parallels the reports of Wilhelm Reich, who found that high humidity affected the flux of orgone energy.

Somewhere between psychokinesis and telepathy is mind control—the ability to impose one's will over another. Wolf Messing, whom we introduced earlier, was purportedly summoned into the presence of Joseph Stalin. Three sets of guards were ordered not to allow him to leave. Stalin then commanded that Messing prove his abilities by getting past the guards. Messing purportedly passed the test without difficulty by influencing the guards to perceive him as the head of the Russian secret service—a man who looked nothing at all like Messing. (Sounds sort of like that Jedi mind trick, doesn't it?)

There are literally hundreds of such cases, and while one, or two, or even a majority might be attributable to mistakes or outright hoaxes, it's highly unlikely that *all* of these stories can be so easily dismissed, especially considering the stature of some of the individuals involved. Consequently, although we have no incontrovertible proof, it seems highly likely that psychokinesis is a reality.

Assuming that reality, it then behooves us to ask whether these are isolated incidents, or if, perhaps, this represents a capability inherent in the species. If the latter, is it an ability that was once more prevalent, but has been lost along the way, or is it a new capability, only beginning to emerge? Do the ubiquitous traditions of gods—human in most respects, but for their preternatural powers—actually reflect a time when psychokinesis was more prevalent? (And, indeed, would not humans, given unlimited power, but still in possession of all their foibles, not behave in precisely the manner portrayed in myth?)

On the other hand, it seems likely that, as mankind attains its final spiritual maturation, such abilities would be handled more responsibly.

Satanism

The religion of Satanism is one based upon the philosophy of Anton Szandor LaVey, as outlined in *The Satanic Bible* and other works. Central is the idea inherited from Nietzsche that an individual must enforce his own meaning on life and rise above the perceived conformity of the masses. There is a libertarian element here; diversity is encouraged, everyone is expected to discover their own sexuality, chart their own personality, and decide their own ambitions in life. In this stress on individuality, Satanism is considered a 'Left-Hand Path' religion.

Nonetheless, we can easily see how this is a religion fully consonant with the philosophies expounded herein. Inherent in Satanism is an open acceptance of the rights of all to pursue the path for which they have incarnated. For some it may be the life of a sexual predator, for another, the prey, and for yet a third, the angst of losing a loved one. The value of these experiences are not ours to judge, for in doing so we judge the prerogatives of The Deity. Moreover, we can see that God entails those behaviors which we mistakenly attribute to Satan, so, indeed those who pray to him under this guise are no more misguided than the Catholic Pope. The proper posture of mankind is one of acceptance, not judgment.

Shape Shifting

The most widely recognized shape-shifting phenomena would have to be the werewolf. The image of Lon Chaney Junior helplessly sprouting hair, claws, and fangs under the influence of the full moon has become so familiar as to be almost comical. For the present discussion, we'll concentrate on some of the more subtle forms of shape shifting, which seem a bit more plausible.

Biologists, neurologists, and psychologists today seem to agree with tribal cultures in recognizing that we all have some animal components within us. Evolution has built up our brain tissue in layers. As a result, the human brain can be seen as being composed of an old reptilian brain, an overlaying mammalian brain, and the most recent and explosive development, the uniquely human cerebral cortex.

Because of this structure, the appetites, drives, and skills of the more primitive components remain latent, and are frequently manifested in subtle (and not so subtle) ways. Reactivating these old and powerful shadows is part of the traditional Shaman's training. This suggests our first class of shape-shifting phenomena, which is really not mystical at all. Through various ritual, hypnotic, and hallucinogenic techniques, it is possible to 'awaken' these repressed animal instincts, temporarily enabling the practitioner to experience the surrounding world through the eyes of another species. Although some might argue that such occurrences are no more than hallucinogenic delusions, the physiological underpinnings suggest that there is no reason to doubt the subjective reality of this class of experiences.

A second variety of shape shifting is more closely related to the concept of astral travel. In this model, the astral body is compelled to separate from the body, but rather than simply roaming free through the astral plane, it is directed to temporarily enter the body of the desired animal. While all of the possible scientific explanations for astral projection would also apply to this phenomenon, it could also be argued that evolutionary animal remnants, when appropriately developed, might contribute an additional affinity to aid in the union.

The final form of shape shifting we'll consider here is of a more physical nature. This comprises changes of appearance, including facial features, body structure, or even hair and eye color. Such changes have been reliably documented in alleged cases of demonic possession. Even if the underlying cause is assumed to be psychological in origin, the physical manifestations cannot be denied. Similarly, the bloodless piercings, firewalking, and related exploits of entranced yogis, as well as the nearly superhuman abilities exhibited by advanced martial artists, both argue that the human body is capable of being controlled in ways and to extents far beyond the limitations commonly attributed to it. Taken together, these phenomena suggest that, directed by a focused and trained mind, the outward appearance is almost certainly capable of alteration.

Yoga/Martial Arts

Yoga and martial arts share many similarities, including their history. Of primary interest in the current context is the well-documented evidence that these disciplines allow practitioners to exert control over their bodies in ways that often border on the paranormal. Examples include firewalking, bloodless body piercings, the extreme slowing of heartbeat and respiration, catching arrows in flight, and so forth. At least one martial arts discipline, known as *Dim Mak*, purports to be able to kill or heal without actual physical contact. (According to some reports, this technique was used against the late Bruce Lee, in retribution for his exposing supposedly secret doctrines to Western audiences.)

The fact that at least some of these claims have been measured and documented under strict laboratory conditions demonstrates that these capabilities exist, and must therefore represent an ability that can be cultivated in at least some individuals. Since the feasibility of many of the models proposed in this book hinge on the ability of the mind to exert these kinds of influences, first over the body itself, and secondly over external events, we can see that at least part of our thesis is thereby proven to be plausible.

These disciplines, however, involve far more than merely being able to control the gross physical body. The goal of this bodily control in Yoga, for example, is to allow the *Kundalini*[1] (universal cosmic energy) to traverse its way through the *chakras*. Kundalini is often described as a coiled storehouse of energy, originating from a reservoir at the base of the spine.

Chakras are energy centers, which can be envisioned as disks, positioned along the spine and at right angles to it. There are generally thought to be seven or eight chakras, the highest of which, the *Sahasrara Chakra*, actually resides above the top of the skull, and corresponds to the state of enlightenment or nirvana.

This concept bears an uncanny resemblance to the high-energy linear particle accelerators used in the study of nuclear physics (see Figure 37).

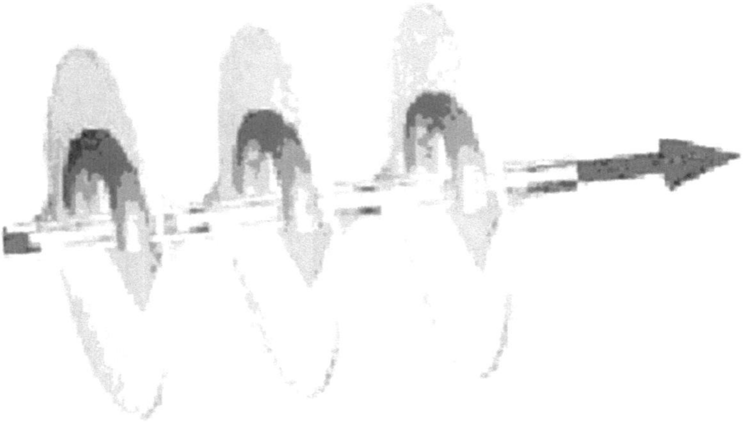

Figure 37

In the particle accelerator, powerful magnetic fields, shown as disks in the illustration, induce an electric field through their centers, which in turn imparts energy to the particles shot down the beam tube. In Figure 38 we see illustrations of eight chakras. These may be envisioned as representing fields, existing at right-angles to the spinal column, analogous to the magnetic fields in a particle accelerator.

[1] Kundalini literally means coiling, like a snake. It refers to a storehouse of universal energy, which in most people lies dormant, in a coiled form near the base of the spine. It is believed to be a source of great power, and can be expanded through certain yogic practices, resulting in the development of paranormal abilities. These include mastery over the elements, the power of leaving and re-entering the body at will, supernormal hearing, weightlessness, enhanced sight, and the control of mind over matter.

Figure 38

The spinal column corresponds to the beam tube, and the kundalini is akin to the accelerated particle beam. The end result is that when we have fully energized the chakras, the consciousness is 'accelerated' into the timeless realm beyond the physical plane.

Although many people are not aware of the fact, most traditional martial arts share a common heritage with yoga, and represent dynamic forms of meditation. Around fifteen hundred years ago, an Indian monk named Bodhidharma traveled to China to spread the practice of Buddhism. Legend has it that when he arrived at the Shaolin temple, he found the monks there in such poor physical condition that they could not meditate effectively. Bodhidharma introduced the monks to five elaborate exercise sets, which simultaneously served to improve their health, while providing a discipline for meditation.

Since both areas of practice have repeatedly demonstrated the ability of the mind to control otherwise autonomous body functions, it is clear that the prospect of controlling aging, or even physical ascension may well be within our reach.

Conclusion

Belief is one of our most powerful assets. Virtually every 'self-help' book on the market shares as a common thread the importance of an underlying belief system. Imagine what you want, convince yourself you can do it, and results will be achieved.

Faith came more easily to our ancestors. Their lives were open to the mystery of the world around them, and their predominant authority figures—shamans, church figures, and so forth—were sources of verification rather than doubt. Over the past century or so, most people have wholeheartedly placed their faith in scientists—the authority figures of our time. As we've seen in the preceding chapters, the doctrine of science has many highly visible successes to its credit, so it's hardly surprising that we tend to put as much faith in it as we do. Science has become an authority that we hardly dare to question.

For many people, science has now become the primary measure against which all things are evaluated, accepted or rejected. As we've seen, however, the ability of scientists to recognize the deeper meaning behind their own discoveries (rituals) is questionable at best. Even in terms of the more mundane aspects of technology, the vast majority of practitioners demonstrate a remarkable propensity to underestimate the scope of what they don't know. (Remember how Lord Kelvin so proudly proclaimed that all of the discoveries of science had already been made!)

With this in mind, the main impetus in preparing this book has been to remove that impediment. No words can convey the essence of reality, but many concepts can effectively prevent us from opening ourselves to gnosis. Chief among these has been the mechanistic doctrines of scientific nihilism. By gathering together those teachings that support the possibility of spiritual content, we have hopefully eradicated that blockage so the mind can become more open to accept the vision of enlightenment.

Atom bombs, photoelectric cells, lasers, and the internet are all undeniable facts of modern life. By illustrating how the underlying mechanisms of these and other discoveries not only *allow* for the possibility of spiritual concepts and paranormal phenomena, but in most cases, *demand* it, I have endeavored to empower the open-minded reader by providing a new, stronger basis for this essential, whole-hearted belief.

Our minds are more powerful and more complex than most people realize. Our capabilities are rarely fully utilized because of self-imposed limitations. (Indeed, it has been said that even the most intelligent among us utilize less than 10% of their mental capacity.) With the ability to believe, you can accomplish nearly anything! The more solid your faith, the more you will accomplish.

Science itself has proven that the mind is intimately connected with physical reality, so the question should not be *can* it be done, but rather, do *you* have the perseverance to master the techniques. This is one aspect that has not changed since the days of our ancestors. The common thread then, as now, was a devotion to the rigorous training necessary to master the self.

Intuition is the dialog between the intellect (consciousness) and the soul, which, in turn, is the part of you that participates in The Universal. It is a form of perception that originates beyond the physical senses. Its purpose is to help you make sound decisions. Throughout this book I have attempted to provide a rational framework within which the cognitive mind can best activate the intuitive channel and allow it to transmit these higher truths unchallenged by rationality.

By internalizing the models presented here you will be better able to choose more wisely, act more compassionately, and hence, will learn more quickly, with greater acceptance and with less pain and suffering.

To conclude, therefore, the meaning of life is:

- to directly participate in the full sublime complexity of The All.
- to have every question answered,
- to attain perfection through experience.

Part VI - Appendices

This section presents an overview of some background incidents that indirectly resulted in the writing of this book. My original intent was to make this the first section, thus providing a chronological introduction. After some consideration, however, I decided that these events would be better understood in light of the concepts presented in the earlier chapters. Also included is glossary of terms.

Appendix A - Evidence of "Something Beyond" (The Ghost Photo)

In this chapter I'll relate one of the personal experiences that served to convince me that there are aspects of the universe beyond the mundane reality normally supported by our senses. This event took place in 1993, shortly after I left my job at Digital Equipment Corporation's Advanced Semiconductor Development lab, and moved with my family to rural Pennsylvania.

Having become increasingly disenchanted with the business of software development, my primary objective was to devote time to writing. Nevertheless, I realized that it would be some time before I'd be able to support myself and my family solely through literary endeavors. Accordingly, I decided to strike off on my own as an independent consultant—an endeavor, I reasoned, that would allow me to work fewer hours, and devote increased time to more gratifying pursuits.

One of the first things I had to do was design a logo for my letterhead. I used a simple drawing program, and printed the graphics out in large scale on a dot-matrix printer. Then I photographed the result using high-contrast techniques, and reduced it in the darkroom. I had used this approach before, and the shrinking process essentially eliminated all the jagged edges left by the dot matrix printer, and resulted in nice camera-ready art.[1]

[1] Remember, this was 1993, and laser printers were prohibitively expensive.

I was in the process of making the final prints, and as I started the timer, my enlarger flashed, and then the bulb went dead. I left the darkroom and drove to the photo store to purchase a new bulb. When I got back I replaced the bulb (under the safelight, which I had left on), put another sheet of paper into the enlarging easel, and continued where I'd left off.

After the paper was exposed, I dropped it into the tray of developing solution, and only then did I notice that I had forgotten the original sheet from the easel. I had inadvertently transferred both, stuck together, into the developer. I could see that the damaged image was rapidly turning black, but really couldn't see any detail under the safelight. Of course, I wouldn't have expected anything intelligible anyway. I slipped both sheets into the fixing bath, then washed and hung them both up to dry. Later, when examining the dry image, I noticed the semblance of a face that sent a chill down my spine (see Figure 39).

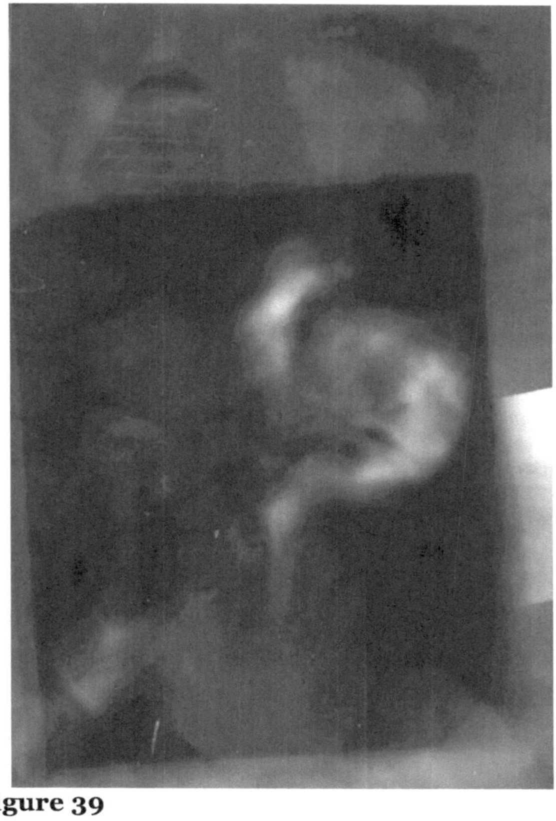

Figure 39

I don't mind confessing that for the remainder of the day I felt reluctant to be alone in the house.

My initial inclination was to attribute the strange image to sloppy darkroom procedures. I surmised that perhaps I had allowed one or more stray negatives to get caught up in the enlarger somewhere. After ripping it down to every last piece, however, I had to abandon that particular theory. Next, I figured that it might have been the result of some stray reflections (especially since the area in the upper left appears to be some kind of screen or grid.) Once again, a careful inspection of the darkroom suggested no possibilities.

Although I could clearly see the border where the second piece of paper had initially stuck to it in the developer, there was still no way to explain the other odd images. (Note, also, what appears to be some kind of figures or symbols near the top, just right of center. I've never been able to make these out. Even with digital enhancement and enlarging, the closest I can come is "LOA 4".)

When my wife returned home later that evening, she immediately sensed that something was wrong. Reluctantly, I showed her the picture. She immediately recognized a vaguely human countenance. As we discussed it, however, we both realized that she was seeing a totally different 'face'. She was even more disturbed when I pointed out the one that I had seen. Later, as we showed it to others, the scenario was repeated. Ultimately, we 'identified' several different faces, though some require a bit more stretch of the imagination than others.

Since the image is so indistinct, I've used Adobe Photoshop™ to emphasize a few of the 'faces'. The results are shown in Figure 40. Since they all tend to overlap, bringing one to the foreground obscures others. There are several more, but these should serve to emphasize what *I* see in the picture.

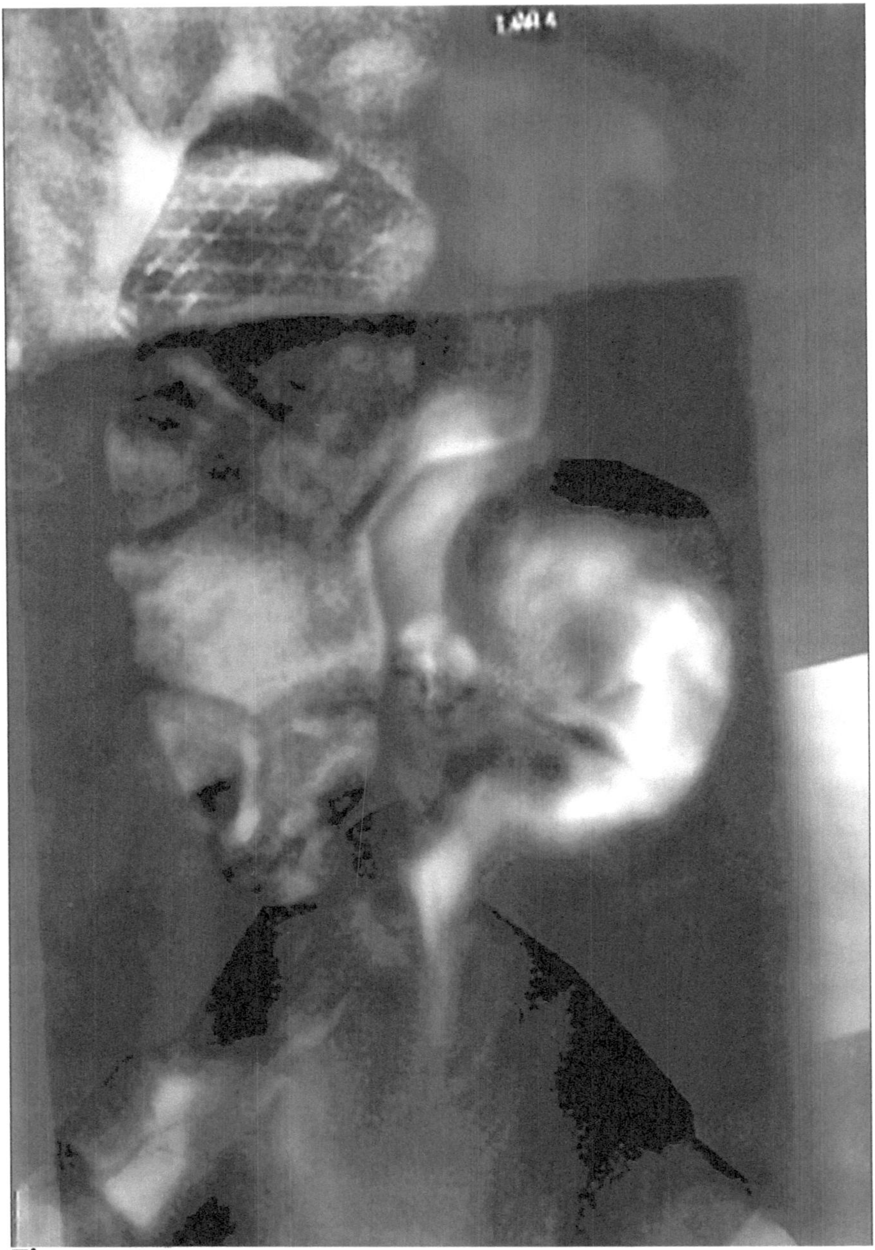

Figure 40

The face I first noticed is shown in Figure 41. The one my wife saw first is separated in Figure 42.

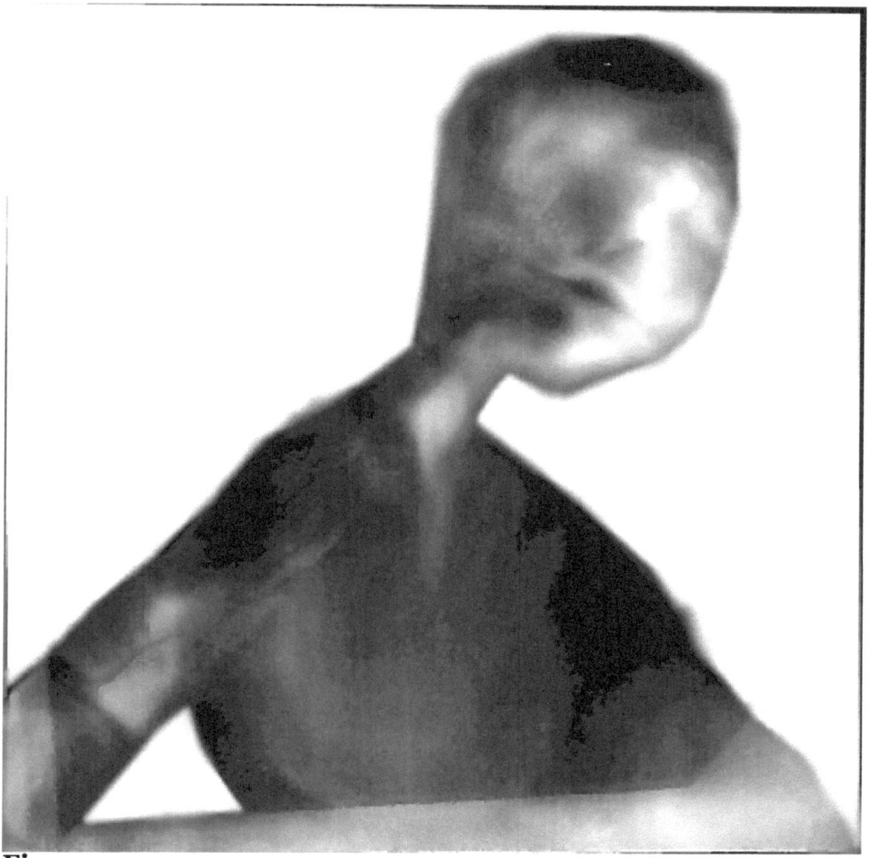

Figure 41

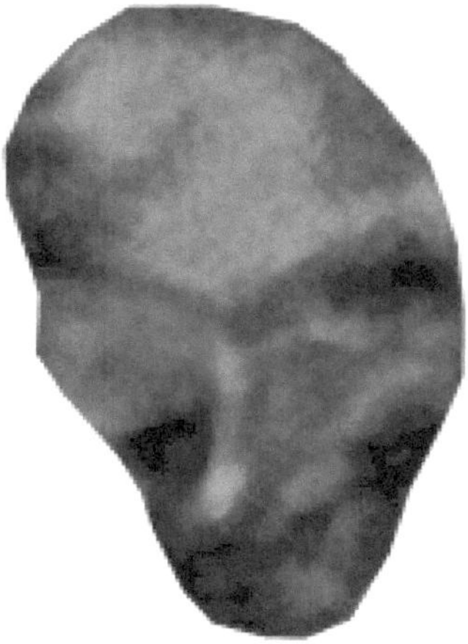

Figure 42

Figure 43 and Figure 44 isolate a few more interesting segments.

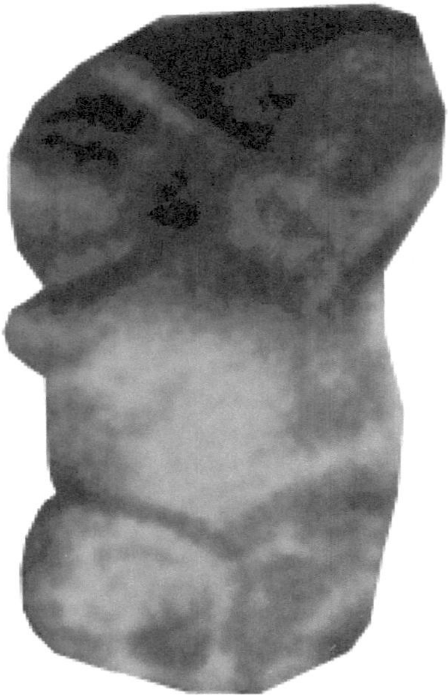

Figure 43

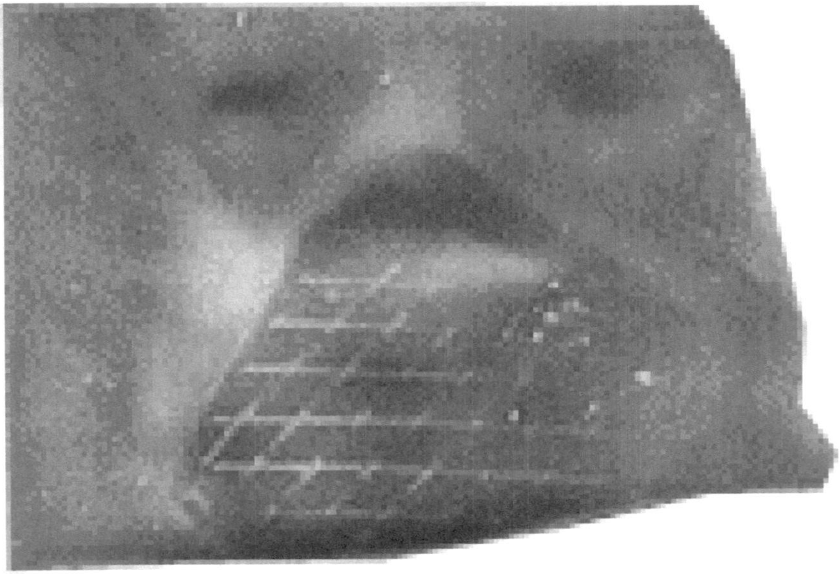

Figure 44

Now, those who are familiar with darkroom techniques and film processing will be aware that high contrast paper should yield primarily black and white images, with very few grays. Moreover, even assuming that I fogged the paper when I left the darkroom to go for a new bulb, fogged paper doesn't result in images as distinctive as these.

Not long afterward, our parish priest claimed that he could feel a malevolent presence in our home, and went so far as to request permission from his bishop to perform an exorcism. Permission was refused. Although we maintain our friendship to this day, he never again set foot in our house.

I've spent the years since this event trying to come up with some meaning for the incident. One abstract possibility I've surmised is that, at the time, I was very much skirting the 'region' between darkness and light. Somehow, perhaps the turmoil of my emotions combined with some freakish trick of space-time to allow the images of what was all around me to become manifest (a rather disturbing thought).

Interestingly, one of the better documented manifestations of psychokenisis is an ability to cause images to form on film.

I have also come to believe that perhaps *everyone* witnesses some seemingly miraculous event that serves to support faith. No one is expected to believe without reason, but likewise, there must be some effort and will associated with faith or there could be no perfecting of the spirit. The problem is that after the initial startled reaction, we tend to just file the matter away and conveniently forget about it as we go about our daily lives. Perhaps this is because it doesn't fit into our materialistically compartmentalized belief system. (This is one of the most unfortunate legacies of secular scientific determinism.)

Just think back and ask yourself whether anything has ever happened to you that was simply inexplicable. I'm convinced that everyone is given the tools necessary to believe, and if they choose to reject them in favor of a more materialistic world-view, then it's not as if they weren't given the chance to make an informed choice. This goes back to the model we've discussed earlier, when we examined the fulfillment of God's nature in the perfection of the human soul.

To the spirit, dwelling as it does in continual witness of the Beatific Vision, there is no need of faith to support what is totally evident. Thus, in order to benefit from the lesson of faith, it is necessary to draw around us a veil of the material; that is, to incarnate into the physical world. Naturally, the tests must be difficult in order to have meaning, but God would set no task that we do not have the capability to accomplish. Hence, everyone is given some clear indication that there is more to existence than the material and the visible. Most, however, tend to obscure the memory and eventually convince themselves that it was merely illusion.

I suspect that in my case the indication is more vivid and persistent than most—I cannot ignore or deny its existence because the picture remains as a constant reminder. This being the case, I'm compelled to wonder—why? Does the clarity of the sign reflect a particularly strong necessity—a stubbornness on my part to cling steadfastly to the material? Is it because the difficulty of my trials requires this constant reminder? Or, perhaps, do extremes of good or evil hinge upon some crucial choice that I may be called upon to make sometime in the future? Perhaps this sign was necessary to compel me finally to write these very words. On these subjects I can do no more than speculate.

As it stands, I keep this strange photo locked away in my desk drawer. Whenever I feel like I'm starting to doubt that there's anything beyond what my five senses can reveal, I only have to look at 'the picture'.

Appendix B - A Ghostly Visitation?

Although the ghost photo described in the previous chapter is certainly the most tangible of my paranormal experiences, it was not the first.

Back in the mid-1970s, my wife and I lived on a 42' 1916 Grover sloop in Gravesend bay, just south of the Verrazano Narrows Bridge in Brooklyn, New York. As we relaxed one evening below decks, there came a horrendous pounding from above that sounded as if someone was banging on the hull with a sledgehammer. I immediately went up to investigate, but as soon as I cleared the companionway hatch, the sound abruptly ceased. I walked around the decks, checking for loose rigging or signs of intruders, but found nothing that could explain the racket. The air and the water were both dead calm.

As soon as I returned to the cabin, however, the pounding returned. Stranger still, my wife, who had remained below, reported that it had never stopped! This time we both went up to look around. I even held my ear against the hull to determine whether, perhaps, it was something below the waterline. I heard nothing.

We both returned to the cabin, somewhat uneasy, but the sound did not return. Several hours later, my parents showed up to deliver the news (we didn't have a telephone) that my mother-in-law had committed suicide earlier that day.

Appendix C - A Conjuring Experience

The following is another true experience, which happened to me in the late 1960s. It is related with greater detail, in fictional format, in my short story "Subtle Shades."[1]

At the time, Marty, my closest friend, and I had been dating two girls who were similarly close friends. When Marty began seeing someone else, the state of affairs threatened both friendships. Lynn (the girl Marty had been seeing), convinced Elaine (my girlfriend) and myself that a love/seduction spell would be just the ticket for remedying the situation.

My recollection of the details of the magical operation are a bit sketchy after all these years. I do recall that we had a curious old book called *The Sixth and Seventh Books of Moses*, which we had found in an old Hassidic bookstore, as well as a copy of *The Black Arts*, by Richard Cavandish.

Elaine lived in an old three-story brownstone in the Borough Park section of Brooklyn. It had a rather deep sub-basement below the walk-in, and that was where we set up to perform our ritual.

We lit candles, burned incense, drew magic circles, and muttered incantations. It was a pretty high-octane spell—all kinds of stuff about anyone coming between the intended lovers being struck down by hellfire, and stuff like that. Suddenly, the basement was filled with a stench the likes of which none of us had ever

[1] J.P. Befumo, "Subtle Shades," in *The Space Between Time*, Philadelphia, Xlibris, 2000, 41.

encountered. We also all saw something . . . it looked sort of like an upside down top, spinning near the ceiling at the far end of the basement. We tried reading a dismissal invocation, but nothing happened. We tried it several more times, then we just ran. We walked the streets for several hours, not wanting to go anywhere near Elaine's house. At last, Lynn and I both called our parents, made some excuses, and went home with Elaine—none of us wanted to be alone.

When we got there, Elaine's mother—an old Sicilian lady—had called in a priest to bless the house. She was frantic about there being "something evil in the house." (In all honesty, she was in the habit of having the house blessed on a regular basis anyhow.)

Later that evening, as we listlessly sat over dinner, Marty called. (Not finding me at home, he'd reasoned I'd probably be with Elaine.) It seemed that his new girlfriend had suddenly been stricken with an acute bout of appendicitis, and had been rushed to the hospital.

After a quilt-ridden conference, we decided that we'd better come clean and let Marty know what had transpired. He thought the whole thing was pretty cool, and wanted to try another spell with us. Reluctantly, we agreed. The next day, when Elaine's mother had gone off to the lampshade factory where she worked, we all met once again in the basement.

We tried a different spell this time, and the effect was even worse. (If you want details, I recommend reading "Subtle Shades.") In any case, we ran around for weeks with the feeling that something was stalking us. The strange part of it is, there was never any real resolution. It ultimately seemed to just fade away.

Recently, when I showed my ghost picture to a psychic of my acquaintance, she immediately said that what I saw in the picture had been following me for a very long time. . . .

I guess the only thing I can say in conclusion is that if I were concocting a story from imagination, I would certainly have come up with a more exciting ending. Nevertheless, I do have a nagging feeling that the final chapter has yet to be played out.

Appendix D: Glossary of Terms

Apodictic — From the field of formal logic, a proposition that is incontrovertibly or demonstrably true as a result of having been deduced systematically, in a stepwise manner, from incontestable foundational principles, in such a way that denying it leads to a contradiction.

Beatific Vision — The immediate and direct knowledge of God, as experienced by the spirits that reside in His presence.

Bell's Theorem — A mathematical proof which demonstrates that if Quantum Theory is correct (which, by all accounts, it is), then nonlocality must be assumed (see *nonlocality*). Bell's Theorem has been called "the most profound discovery in the history of science."

Big Bang — The prevailing theory of the beginning of the universe, in which at the moment of creation, all matter exploded from a volumeless point, and time, matter, energy simultaneously came into being.

Bilocation — A theological concept which holds that certain saints had the ability to appear physically in more than one place at the same time.

Bohm, David — David Bohm was one of the world's greatest quantum mechanical physicists and philosophers of the century. Bohm envisioned a world of interconnection and interdependence; of direct and instantaneous communication, in which we would learn to harness the energies of compassion. He believed in a world that is meaningful, clear, intelligent and spiritual, and in which implicit order is expressed as a living force in our lives.

Calculus – The branch of mathematics that deals with limits, infinitesimals, and the integration and differentiation of functions of one or more variables.

Chakras - According to Tantric philosophy, chakras are points of energy in the human body, orthogonal to the spine. Each chakra has certain qualities and associations. The practice of yoga is used to gradually elevate the consciousness through the seven chakras, the highest of which corresponds to nirvana.

Christ-Consciousness – The Essence of the regenerative and reconciliatory aspect of the force of pure being, and the Soul Essence within each one of us, representing the unfolding energy of wholeness in human nature.

Classical Mechanics – The branch of science or engineering dealing with macroscopic objects that are large compared to the size of an energy quanta, moving at velocities that are slow with respect to the speed of light.

Classical Relativity – the mathematical formulations used to translate the dynamics of objects within frames of reference in motion with respect to each other, but moving slowly compared with the speed of light.

Closed Friedmann-Lemaître theory – a theory about the ultimate fate of the universe, which holds that the universe is a spherically curved continuum, destined ultimately to collapse back in on itself.

Common-nature – The archetype of the human soul, the *Christ-Consciousness.*

Complex Mathematics – The branch of mathematics that deals with the complex plane; that is, the set of all mathematical entities which have a component that exists in the so-called 'imaginary' plane (see *imaginary numbers*).

Copenhagen Interpretation – The most conservative interpretation of quantum mechanics, which holds that QM is a complete theory, even though it can not predict anything about the details of specific interactions.

Delayed double-slit experiment – An experiment that appears to have proven that, at the quantum level at least, particles can somehow 'know' ahead of time how the experiment is going to be carried out, or, alternatively, that an action performed in the present can alter the events of the past.

Diffraction – A change in the directions and intensities of a group of waves after passing through an aperture whose dimension is small compared to the wavelength of the waves.

Double-slit experiment – A classical optics experiment used to demonstrate the wave nature of light. Light waves are made to pass through two thin slits, and shine on a screen. The resulting bands of light and darkness identify the places where the waves reinforce or cancel each other out.

Einstein-deSitter theory – a theory about the ultimate fate of the universe, which holds that the universe is essentially flat, and is fated to expand forever.

Electromagnetic induction – The phenomenon, discovered by Michael Faraday in 1831, by which a changing magnetic field can induce an electrical current.

EPR paper – An effort by Einstein, Podolsky and Rosen to demonstrate that Quantum Theory is wrong. They set out to show that, carried to its logical conclusion, Quantum Theory led to an absurdity—the concept of nonlocality. Later, John Bell demonstrated that in fact, Quantum Theory is correct, and nonlocality is real. (See *Nonlocality*.)

Everett, Hugh — Physicist and author of the 'Many-worlds' interpretation of Quantum Theory.

Faraday, Michael — Discoverer of electromagnetic induction, electromagnetic rotations, the magneto-optical effect, diamagnetism, and field theory.

Flatland — A hypothetical two-dimensional universe used to help students of physics come to grips with higher orders of dimensionality.

Gnosis — The intuitive apprehension of spiritual truth.

Gnostics — A group of early Christian sects that emphasized an esoteric form of knowledge achieved through direct intuition.

General Relativity — A theory of gravitation, formulated by Albert Einstein, which equates all matter, energy, and gravitation with irregularities in the fabric of the space-time continuum.

Hiesenberg Uncertainty Principle — A mathematical formulation of the physical law that observing a system must always necessarily change it.

Hilbert spaces — A Hilbert Space is a nonlinear vector space that can have an infinite number of dimensions. The dimensions of the Hilbert Space have nothing to do with physical dimensions as we commonly understand them, but mathematically represent the collection of all analytic functions.

Hologram — An imaging technique in which a split laser beam is used to record a three-dimensional representation of a solid object. The resulting recording can be broken into arbitrarily small sections, and each will still project the entirety of the original image, though at lower resolution.

Imaginary numbers — A numbering system based around the square root of negative one.

Interference pattern — The pattern that results when two sets of waves interact in such a way that when two crests or two troughs line up, they reinforce each other, but when a crest aligns with a trough, they cancel each other out.

Karma — An impersonal process, or natural law, through which the soul is incarnated into situations in which needed lessons may be learned.

Klein Bottle — A closed, one-sided, non-orientable surface which exists in four-dimensional space.

Light-year — a unit of measurement corresponding to the distance light travels in one year. One light-year is equal to approximately 9,500,000,000,000 kilometers.

Loa — A group of spirits worshiped by practitioners of the Voodun religion. Loa means "mystery" in the Yoruba language.

Lorentz Transformation — A mathematical formula of classical relativity used to transform coordinates from one moving system to another.

Many-world theory — An interpretation of Quantum Theory which holds that every possible combination of events exists as a set of mutually orthogonal universes.

Maxwell's Equations — Four equations, developed by James Clerk Maxwell, which together fully explain the physics of electromagnetic radiation.

Metaphysics — The branch of philosophy that addresses questions pertaining to the ultimate composition of reality.

Metastable state — A quantum state in certain crystalline materials in which excited electrons will gather before returning to their ground state.

Michelson-Morley experiment — One of history's most famous experimental failures, this experiment sought to measure the speed of the earth through the ephemeral aether. The fact that no such effect was found led Albert Einstein to postulate a constant speed of light. This assumption ultimately resulted in the formulation of his theory of Special Relativity.

Möbius strip — A topological anomaly, the Möbius strip is a two-dimensional object that has only one edge and only one side. It was discovered in 1858 by German mathematician and astronomer August Ferdinand Möbius.

Netwon's Laws — Three axioms that comprise the laws of motion, and which form the foundation of classical mechanics.

Nonlocality — The theory which holds that distant quantum particles exert influences on each other at seemingly instantaneous rates. This implies that distant points may be intimately connected in ways that cannot be perceived from our three-dimensional perspective.

Nutrino — A chargeless, massless particle thought to permeate all of space.

Open Friedmann-Lemaître theory — a theory about the ultimate fate of the universe, which predicts a hyperbolically curved space-time continuum that is destined to expand forever,

Orthogonal — Relating to, or composed of right angels. Typically, the intersection of two orthogonal objects is an object of one less dimension than the intersecting objects themselves. Hence, two orthogonal lines intersect at a point; two orthogonal planes intersect at a line, and so forth.

Particle —- An object whose volume is sufficiently small within a particular context that it can be considered negligible.

Process Philosophy – A world-view developed by Alfred North Whitehead, which holds that reality is essentially episodic in nature, and that only the mind of the observer imparts to it a sense of continuity.

Quantum Linearity – A concept related to the many-worlds interpretation of Quantum Theory, which holds that the multiple worlds can interfere with each other from the perspective of an external observer, but the interfering worlds cannot influence one another.

Quantum mechanics – The theory that describes the behavior of matter and energy at the smallest distance scales.

Quark – A group of elementary particles having fractional electrical charges, and considered the building blocks of all hadrons (subatomic particles found in the nucleus of atoms).

Right angle – An angle of precisely ninety degrees, as in the corners of a square or rectangle.

Shango – Shango is the god of thunder in Yoruba legend, and one of the central figures in their spiritual lore.

Schrödinger equation – The central formulation of modern Quantum Theory.

Shroedinger's Cat – A 'thought experiment' used to convey some of the principle concepts of Quantum Theory.

Singularity – A point mass with absolutely no extent in space, that is, no volume whatsoever, but existing at a single point.

Space-time continuum – A four-dimensional matrix consisting of our three spatial dimensions, plus time, arranged in such a manner that for any two points, regardless of how close together they are, a third point can always be found between them. The significance of this formulation is that time is treated in a manner identical with the spatial dimensions.

Special Relativity – A theory describing the dynamics of particles moving at velocities close to that of light. Among the predictions of this theory are the equivalence of mass and energy ($E=mc^2$), time dilation, and the space-time continuum.

Super determinism – The notion, seemingly implied by Special Relativity, that everything--past, present, and future--'already' exists in some sense, and as a result, free will is merely an illusion. Super determinism is negated by the many-worlds theory, and concepts such as process philosophy, which have *every possible outcome* pre established, with free will emerging as the means by which will is exerted.

Topology – The study of the properties of surfaces not normally affected by changes in size or shape.

Universal Spirit – The combined souls of all sentient creatures, which together represent that aspect of God associated with agape love.

Vector spaces – The fundamental concept in linear algebra, it is a generalization of the set of all geometrical vectors and is used throughout modern mathematics.

Wave – A periodic, propagating energy form distributed through space.

Wave function – A mathematical function used in quantum mechanics to describe the probability associated with a particle or group of particles.

Wheeler, John – A renowned mathematician and physicist who described the universe as essentially participatory; that is, the universe does not exist until a human mind observes it.

Whitehead, Alfred North – A philosopher and mathematician who developed the concept of Process Philosophy.

Thank You For Reading

I'd like to thank you for expending the effort to read this book. I sincerely hope that it in some small way contributes to making your life more complete, your load lighter, and your vision more perspicacious.

As an independent author, word of mouth is my most effective advertising. A review at your favorite online bookseller, or mention on your website, on facebook, twitter, or just plain old fashioned word of mouth takes only a few minutes, and really helps a lot. Your opinion is valued!

Joe Befumo

See my other written works at:
http://www.PleasantMountPress.com/jbefumo
Feel free to email me: jbefumo@PleasantMountPress.com

Made in United States
Troutdale, OR
11/03/2026

56852530R00156